ConcepTests

to accompany

CALCULUS

Fourth Edition

Deborah Hughes-Hallett
University of Arizona

Andrew M. Gleason
Harvard University

William G. McCallum
University of Arizona

Prepared by:
Scott Pilzer, College of the Redwoods
Maria Robinson, Seattle University
David Lomen, University of Arizona
Daniel Flath, Macalester College
Brigitte Lahme, Sonoma State University
Jerry Morris, Sonoma State University
Ann Davidian, General Douglas MacArthur High School
David Lovelock, University of Arizona

WILEY
JOHN WILEY & SONS, INC.

Cover Photo Credit: ©Pete Turner/The Image Bank/Getty Images

To order books or for customer service call 1-800-CALL-WILEY (225-5945).

ISBN 0-471-65999-1

Printed in the United States of America.

10 9 8 7 6 5 4 3 2 1

Printed and bound by Malloy Lithographing, Inc.

INTRODUCTION TO CONCEPTESTS

Over the past two decades, many faculty have found ways to improve student learning by making better, more active, use of class time. One of the most effective methods—the use of ConcepTests and peer instruction—was developed by Eric Mazur of Harvard. His pioneering work in physics has been successfully replicated in other departments, and its efficacy is clearly supported by data. Information is available under "Peer Instruction" at http://galileo.harvard.edu.

This collection of ConcepTests for Calculus represents the first time the method has been made widely available for mathematics. Evaluation suggests that ConcepTests are as effective in calculus as they are in physics.

What are ConcepTests and How are They Used?

Our ConcepTests are questions designed to promote the discussion and learning of mathematical concepts during a calculus class. The questions are usually conceptual, often multiple choice or True/False, with some free response questions as well. They are used as an aid in promoting student involvement in discussing mathematical concepts rather than as a method of testing students for a grade. (Some questions have more than one correct answer.)

Because of the variety of forms of these questions, teachers can use them in a manner that fits comfortably with their teaching style. Three ways we have used them are:

1. As an introduction to a topic. This works especially well if the topic is closely tied to previous lessons or is something with which most students have some familiarity.

2. After presentation of a specific topic by the teacher. Here the ConcepTest may be used to see if the students have grasped the concept, or if the topic needs more discussion or examples.

3. As a review of material that has been thoroughly discussed.

Instructors usually display the ConcepTest using an overhead projector, or distributing a copy to each student. Students are then given short time (one to four minutes, depending on the question) to think about the question and then vote for the answer they think is correct. Providing almost all do not vote for the correct answer, the students are then given a few minutes to discuss the Concep Test with adjacent students and then are given another chance to vote on the correct answer. Besides determining the correct answer, students are also to develop reasons to support their answer. The instructor then discusses the correct answer (or answers) and has students present their reasons.

In this collection, ConcepTests are accompanied by answers and comments, which give suggestions for follow-up activities.

Why do ConcepTests Work?

Calculus instructors often strive to involve students in classroom discussions. However, it is not uncommon for their efforts to be met by silence as students wait for the instructor to answer his or her own questions. ConcepTests work because they remove the barriers to student response. At the outset, instructor and students both know that there will be several minutes to grapple with the question, so there is no pressure on the instructor to answer too quickly. The responses are essentially anonymous, so students can comfortably put forth tentative ideas. If the students are lost or unsure, they have the opportunity to talk things over with their classmates and to revise their first impressions. The give-and-take helps both weak and strong students and, even if it does not generate a correct answer, often generates excellent questions from the students to the teacher.

Regular use of ConcepTests ensures that students first grapple with difficult mathematical ideas in class, where help is available, rather than on homework, when help is not available. Having wrestled with ideas in class, students are much more likely to be able to tackle problems actively on their own.

Conceptual Questions for Use Outside of Class

For instructors who want to assign conceptual questions as homework, or students who want to practice on their own, we suggest using the "Check Your Understanding" problems at the end of each chapter of the text.

Evaluation of ConcepTests

In his article "Peer Instructions in Physics & Mathematics"; *Primus*, Volume XI, Number 2, June 2001, pp. 185–192, Scott Pilzer describes Eric Mazur's development of ConcepTests and Mazur's evaluation data. Scott Pilzer, who taught with ConcepTests in physics before writing them for Calculus, gathered similar evaluation data for mathematics. He found that at the start of the subsequent semester, students taught Calculus I using ConcepTests and peer instruction performed much better on conceptual questions and somewhat better on standard problems than those who had been taught standard lectures. See Table 1.

Table 1 *Performance of Calculus I students at the start of Calculus II*

	Conceptual problems	Standard problems
Taught with ConcepTests	73%	63%
Traditional Lecture	17%	54%

These dramatic results mirror the results found in physics: a large increase in conceptual understanding in addition to an increase in standard computational problem solving ability.

What Kinds of Classes Benefit from ConcepTests?

While originally developed for use in a large classroom, we find ConcepTests equally effective in small or moderate sized classes, where they promote "active" or "discovery" student learning. The ensuing discussion greatly increases the students' familiarity with the subject and helps them formulate mathematical ideas in their own words. This increases the students' base of knowledge and enthusiasm for mathematics. In his article, Scott Pilzer describes how this enthusiasm translated into more mathematics majors.

We encourage you to try teaching with ConcepTests and to take part in developing expertise with a pedagogy that promises to be as productive for mathematics as it has been for physics.

David Lomen
Daniel Flath
Deborah Hughes Hallett
for
The Calculus Consortium

CONTENTS

Chapter One

ConcepTests and Answers and Comments for Section 1.1

1. Which of the following functions has its domain identical with its range?

   (a) $f(x) = x^2$
   (b) $g(x) = \sqrt{x}$
   (c) $h(x) = x^3$
   (d) $i(x) = |x|$

   ANSWER:

   (b) and (c). For $g(x) = \sqrt{x}$, the domain and range are all nonnegative numbers, and for $h(x) = x^3$, the domain and range are all real numbers.

   COMMENT:

   It is worth considering the domain and range for all choices.

2. Which of the following functions have identical domains?

   (a) $f(x) = x^2$
   (b) $g(x) = 1/\sqrt{x}$
   (c) $h(x) = x^3$
   (d) $i(x) = |x|$
   (e) $j(x) = \ln x$

   ANSWER:

   (a), (c), and (d) give functions which have the domain of all real numbers, while (b) and (e) give functions whose domain consists of all positive numbers.

   COMMENT:

   For alternate choices, replace (a) with $1/x^2$ and (e) with $\ln|x|$.

3. Which of the following functions have identical ranges?

   (a) $f(x) = x^2$
   (b) $g(x) = 1/\sqrt{x}$
   (c) $h(x) = x^3$
   (d) $i(x) = |x|$
   (e) $j(x) = \ln x$

   ANSWER:

   (a) and (d) give functions which have the identical range of nonnegative numbers. (c) and (e) also give functions which have the same range—all real numbers.

   COMMENT:

   You could have students find a function with the same range as (b) : x^3/x for example.

4. Which of the graphs is that of $y = x^2/x$?

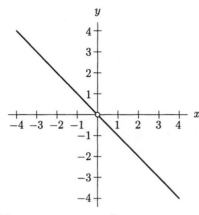

(b)

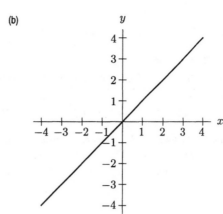

(c)

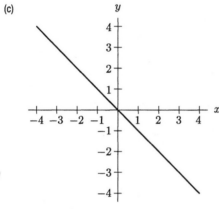

(d)

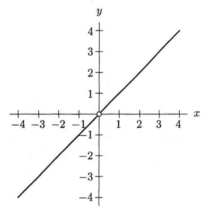

ANSWER:

(d). x^2/x is not defined at $x = 0$.

COMMENT:

This is where you could emphasize that the domain is a crucial part of the definition of a function.

Follow-up Question. Give another formula for the function $y = x^2/x$.

Answer. $y = \begin{cases} x, & x \neq 0 \\ \text{undefined}, & x = 0. \end{cases}$

5. The slope of the line connecting the points $(1, 4)$ and $(3, 8)$ is

(a) $-\dfrac{1}{2}$

(b) -2

(c) $\dfrac{1}{2}$

(d) 2

ANSWER:

(d). $\dfrac{\text{rise}}{\text{run}} = \dfrac{8 - 4}{3 - 1} = \dfrac{4}{2} = 2.$

COMMENT:

You might point out in finding slopes, the order of the points in the ratio $(y_2 - y_1)/(x_2 - x_1)$ is immaterial.

6. Put the following linear functions in order of increasing slope.

   (a) $y = \pi x + 9$
   (b) $y = 3x + 1$
   (c) $y = -10x$
   (d) $y = x$
   (e) $y = \dfrac{x}{10} + 7$
   (f) $y = -100$

   ANSWER:

   (c), (f), (e), (d), (b), (a). In order to put the lines in the correct order, consider the slope of each function.

   COMMENT:

   This question was used as an elimination question in a classroom session modeled after "Who Wants to be a Millionaire?", replacing "Millionaire" by "Mathematician".

7. List the lines in the figure below in the order of increasing slope. (The graphs are shown in identical windows.)

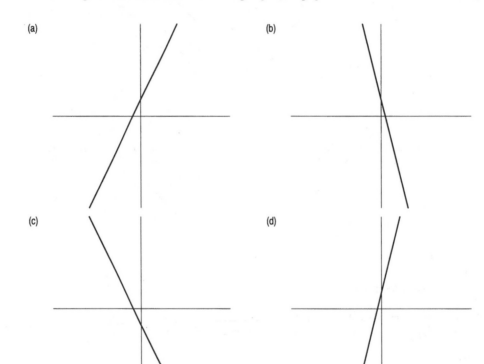

   ANSWER:

   (b), (c), (a), (d).

   COMMENT:

   You may want to point out the difference between slope (numerical values) and steepness (absolute value of the slope).

8. Which of the following lines have the same slope?

   (a) $y = 3x + 2$
   (b) $3y = 9x + 4$
   (c) $3y = 2x + 6$
   (d) $2y = 6x + 4$

   ANSWER:

   (a), (b), and (d). Solving for y in the last three choices gives $y = 3x + \frac{4}{3}$, $y = \frac{2}{3}x + 2$, and $y = 3x + 2$. Thus (a), (b), and (d) have the same slope, 3.

   COMMENT:

   You could point out that (a) and (d) are the same line.

9. Which of the following tables could represent linear functions?

(a)	x	$f(x)$	(b)	x	$g(x)$	(c)	x	$h(x)$	(d)	x	$j(x)$
	1	1		1	-12		1	10		1	12
	2	2		2	-9		2	9		2	14
	3	4		3	-6		4	6		4	16
	4	8		4	-3		8	3		8	18

   ANSWER:

   (b). This is the only table with a constant difference, -3, for the same increase in x. Therefore, (b) is the only one representing a linear function.

   COMMENT:

   You could point out that (d) fails to be a linear function because the x values do not increase by a constant amount even though the function values do.

10. Which of the following lines have the same vertical intercept?

    (a) $y = 3x + 2$
    (b) $3y = 9x + 4$
    (c) $3y = 2x + 6$
    (d) $2y = 6x + 4$

    ANSWER:

    (a), (c), and (d). Solving for y in the last three choices gives $y = 3x + \frac{4}{3}$, $y = \frac{2}{3}x + 2$, and $y = 3x + 2$. Thus (a), (c), and (d) have the same vertical intercept, 2.

    COMMENT:

    You could graph these four equations and then ask about the intercept.

11. Every line has a vertical intercept.

    (a) True
    (b) False

    ANSWER:

    (b). The line $x = a$, where $a \neq 0$, does not have a vertical intercept.

    COMMENT:

    Note that every non-vertical line has a vertical intercept.

12. Every line has a horizontal intercept.

    (a) True
    (b) False

    ANSWER:

    (b). The line $y = a$ where $a \neq 0$, does not have a horizontal intercept.

    COMMENT:

    Note that all lines with a nonzero slope have a horizontal intercept.

13. Every line has both a horizontal intercept and a vertical intercept.

    (a) True
    (b) False

    ANSWER:

    (b). Either a horizontal line or a vertical line, excluding the horizontal and vertical axes, provides a counterexample.

    COMMENT:

    Note that all lines of the form $y = mx + b$, where $m \neq 0$, have both types of intercepts.

14. Every non-horizontal line must have at most one horizontal intercept.

    (a) True
    (b) False

    ANSWER:

    (a). Lines of the form $y = mx + b$, where $m \neq 0$, have $-\dfrac{b}{m}$ as the horizontal intercept, while vertical lines have the form $x = a$, where a is the horizontal intercept.

    COMMENT:

    Ask the students what this means geometrically.

15. The graph in Figure 1.1 is a representation of which of the following functions?

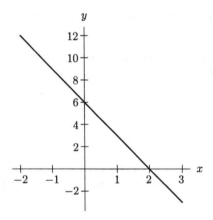

Figure 1.1

    (a) $y = 6x + 6$
    (b) $y = -3x + 6$
    (c) $y = -3x + 2$
    (d) $y = 6x - 2$

    ANSWER:

    (b). The line has slope -3 and y-intercept 6.

    COMMENT:

    Other methods of reasoning could be used. For example, the line shown has a negative slope, which eliminates choices (a) and (d). The y-intercepts for choices (b) and (c) are 6 and 2, respectively. From the graph the y-intercept is 6.

16. The graph in Figure 1.2 is a representation of which of the following functions?

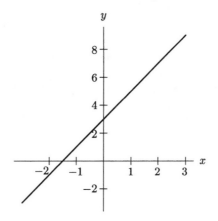

Figure 1.2

(a) $y = 3x - 2$
(b) $y = 2x + 1.5$
(c) $y = 2x + 3$
(d) $2y = 6x - 3$

ANSWER:

(c). The line has slope 2 and y-intercept 3.

COMMENT:

Note that the y-intercept on the graph is positive, which eliminates choices (a) and (d). The lines given in choices (b) and (c) have the same slope, so the choice will depend on the y-intercept, which appears to be 3, not 1.5.

17. Consider the function $f(x) = x^2 + 2x$. Give an equation of a line that intersects the graph of this function

(a) Twice
(b) Once
(c) Never

ANSWER:

(a) There are many answers here. Any horizontal line of the form $y = a$, where $a > -1$.
(b) There are many answers here. One example is $y = -1$.
(c) There are many answers here. Any horizontal line of the form $y = a$, where $a < -1$.

COMMENT:

If you consider this question graphically, then have your students draw non-horizontal lines that meet the requirements for (a) and (b). This could be a way to introduce the idea of tangent lines.

18. Consider the function $f(x) = 2\sin x$. Give an equation of a line that intersects the graph of this function

(a) Once   (b) Never   (c) An infinite number of times

ANSWER:

(a) There are many answers here. For example any vertical line will do.
(b) Any horizontal line of the form $y = n$ where $|n| > 2$.
(c) Any horizontal line of the form $y = n$ where $-2 \leq n \leq 2$.

COMMENT:

Follow-up Question. Draw a line that intersects the graph of this function

(a) Twice
(b) Three times
(c) Four times

You can also ask if it is possible for a line which intersects the curve at an intercept to intersect the curve an even number of times.

19. Which of the following functions is *not* increasing?

(a) The elevation of a river as a function of distance from its mouth

(b) The length of a single strand of hair as a function of time

(c) The height of a person from age 0 to age 80

(d) The height of a redwood tree

ANSWER:

(c). In general, people stop growing when they are young adults and, before they are 80, they begin to lose height.

COMMENT:

This question expands a student's idea of a function. You could ask students to supply some more functions that are increasing.

20. Which of the following lines represent decreasing functions?

    (a) $x + y = 2$ (b) $x - y = -2$ (c) $2x - 3y = 6$ (d) $2x + 3y = -6$

ANSWER:

(a) and (d). Lines with negative slopes are decreasing functions. (a) has a slope of -1 and (d) has a slope of $-\dfrac{2}{3}$, and thus are decreasing functions.

COMMENT:

These equations could also be graphed to show the slopes geometrically.

21. Which of the graphs does not represent y as a function of x?

(a)
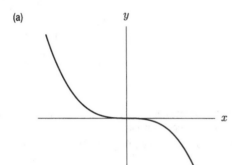

(b)
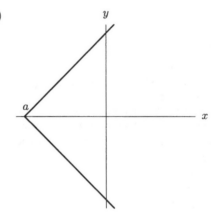

(c)
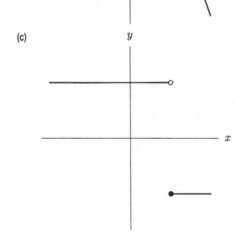

(d)
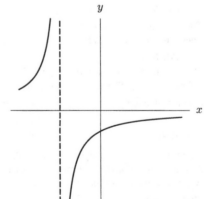

ANSWER:

(b). For $x > a$ there are two function values corresponding to the same value of x.

COMMENT:

   It may be worth noting that we also are not certain about choice (d). All we know is that it looks like a function for the range shown.

22. All linear functions are examples of direct proportionality.

    (a) True

    (b) False

        ANSWER:

        (b). Any linear function whose graph does not pass through the origin is not an example of direct proportionality.

        COMMENT:

        Students should try to find examples as well as counterexamples anytime a definition is introduced.

23. Which of the following graphs represent y as directly proportional to x?

(a)

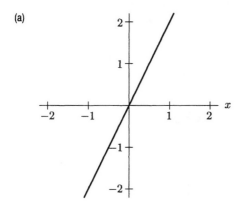

Figure 1.3

(b)

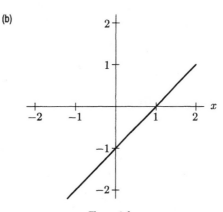

Figure 1.4

(c)

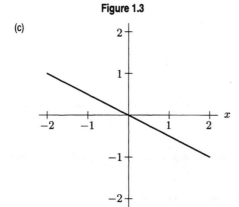

Figure 1.5

(d)

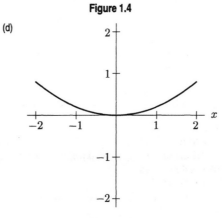

Figure 1.6

        ANSWER:

        (a) and (c). If y is directly proportional to x, then $y = kx$, where k is a constant. (a) and (c) are graphs of such equations.

        COMMENT:

        Note that graphs representing y as directly proportional to x are lines through the origin. Students should recognize the graphical properties of y being directly proportional to x. Notice that (d) could be a representation of y being directly proportional to some **even power** of x.

24. Which of the graphs represents the position of an object that is slowing down?

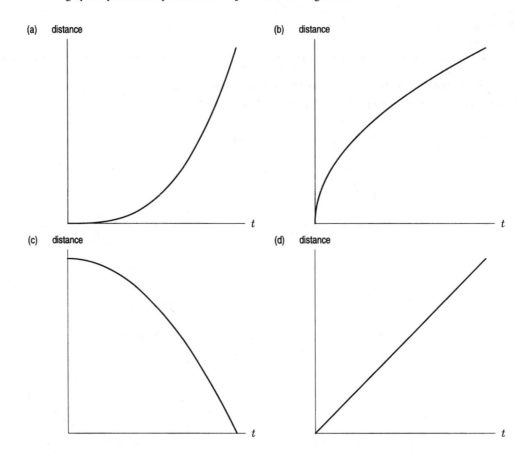

ANSWER:

(b). If the object is slowing down, then the changes in distance over various time intervals of the same length will decrease as time increases.

COMMENT:

You could have your students describe how the object is traveling in the other choices.

25. Which of the graphs represents the position of an object that is speeding up and then slowing down?

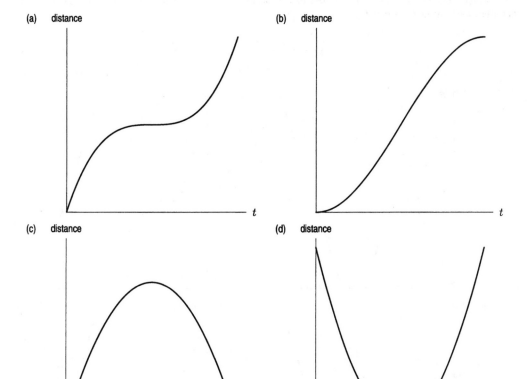

(a)   distance   t

(b)   distance   t

(c)   distance   t

(d)   distance   t

ANSWER:
(b). The graph has a positive slope everywhere, which is increasing for t near zero and decreasing for larger times.
COMMENT:
You could have students describe how the object is traveling in the other choices.

26. There are three poles spaced 10 meters apart as shown in Figure 1.7. Joey walks from pole C to pole B, stands there for a short time, then runs to pole C, stands there for a short time and then jogs to pole A. Which of the following graphs describes Joey's distance from pole A?

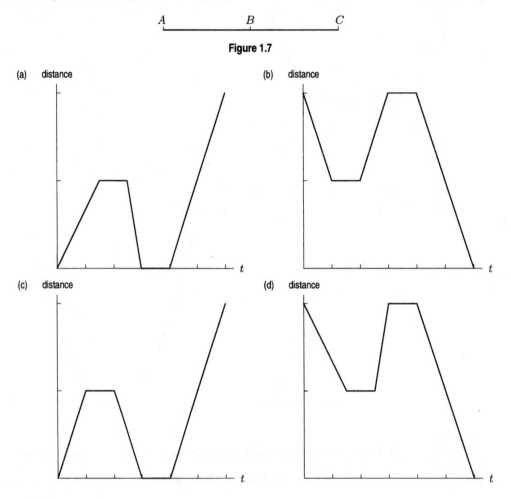

Figure 1.7

ANSWER:

(d). The graph describes the distance from pole A, so it cannot start at the origin. This eliminates choices (a) and (c). Joey will spend more time walking between poles than running. This eliminates (b).

COMMENT:

You can also reason using slopes ($|\text{slope}| = $ speed) as follows: In the first interval he is walking, then he is going three times as fast (running), and finally he is jogging (half the speed of running).

Follow-up Question. Give a scenario that describes the remaining graphs.

ConcepTests and Answers and Comments for Section 1.2

1. The graph of a function is either concave up or concave down.

   (a) True
   (b) False

   ANSWER:

   (b). A function can change concavity or can be a straight line and have no concavity.

   COMMENT:

   Show the different possibilities.

For problems 2–5, use the graphs given below.

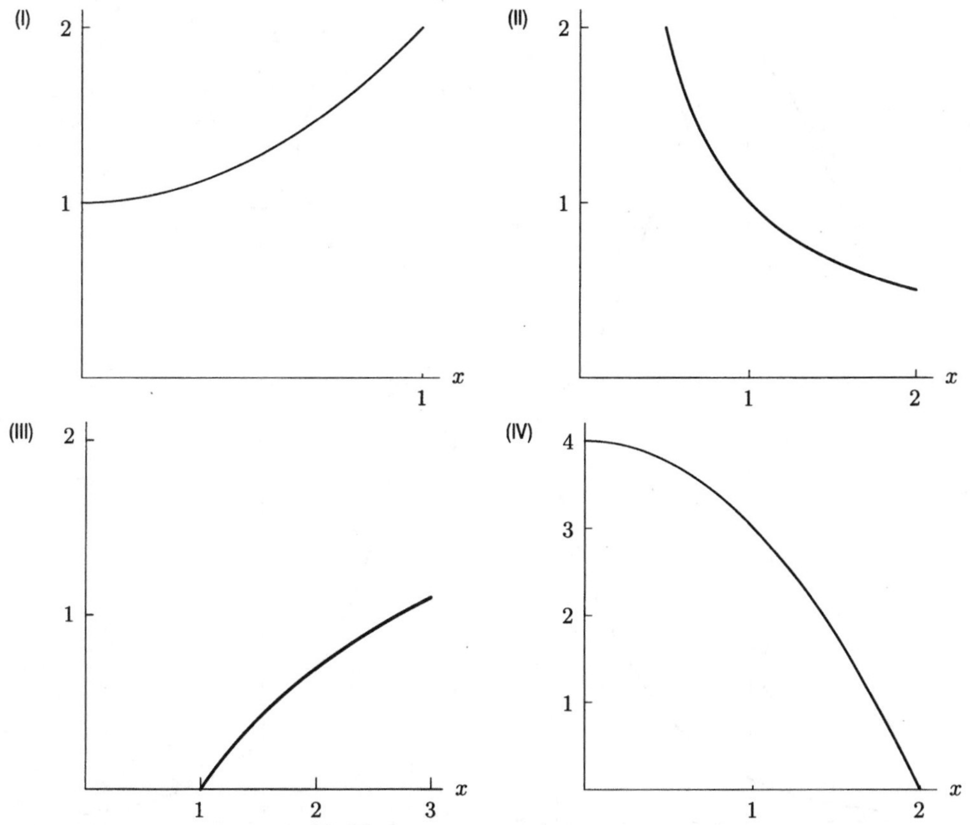

2. Which graph shows a function that is increasing and concave down?
    ANSWER:
    (III)
    COMMENT:
    Follow-up Question. Is this still true if the graph is shifted up one unit?

3. Which graph shows a function that is increasing and concave up?
    ANSWER:
    (I)
    COMMENT:
    Follow-up Question. Is this still true if the graph is reflected across the y-axis?

4. Which graph shows a function that is decreasing and concave down?
    ANSWER:
    (IV)
    COMMENT:
    Follow-up Question. Is this still true if the graph is reflected across the x-axis?

5. Which graph shows a function that is decreasing and concave up?
    ANSWER:
    (II)
    COMMENT:
    Follow-up Question. Is this still true if the vertical distance from the origin is doubled at every point on the graph?

6. Which of the graphs is that of $y = 2^x$?

(a)
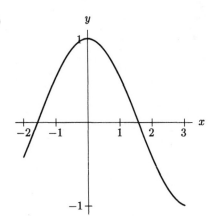

(b)
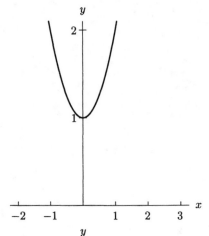

(c)
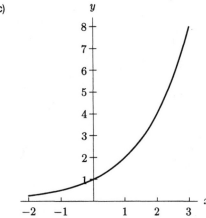

(d)
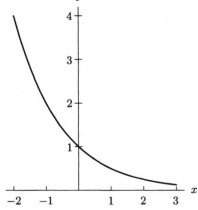

ANSWER:
(c)
COMMENT:
You could have students explain why each of the other choices is not appropriate.

7. Which of the graphs is that of $y = 2^{-x}$?

(a)

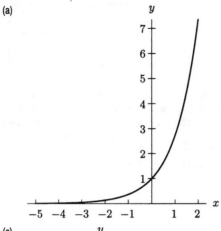

(b)

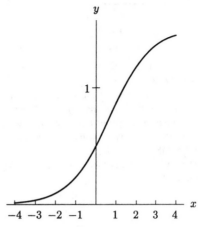

(c)

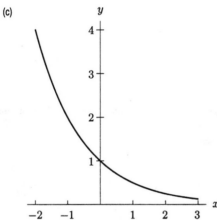

(d)

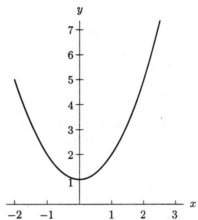

ANSWER:

(c)

COMMENT:

You could have students explain why each of the other choices is not appropriate.

8. Every exponential function has a vertical intercept.

   (a) True
   (b) False

   ANSWER:

   (a). Exponential functions have the form ab^x, where $b > 0$, but $b \neq 1$. The vertical intercept is $a \neq 0$.

   COMMENT:

   Show this graphically—include positive and negative values of a and values of b which are between 0 and 1 as well as values of b which are greater than 1.

9. Every exponential function has a horizontal intercept.

   (a) True
   (b) False

   ANSWER:

   (b). For example, $y = 2^x$ has no horizontal intercept.

   COMMENT:

   Point out that no exponential function has a horizontal intercept. Exponential functions must be in the form $y = ab^x$ where $a \neq 0$ and either $0 < b < 1$ or $b > 1$.

10. Which of the following tables could represent an exponential function?

(a)	x	$f(x)$	(b)	x	$g(x)$	(c)	x	$h(x)$	(d)	x	$k(x)$
	1	1/16		1	9		1	1		1	10
	2	1/8		2	−3		2	4		2	5
	3	1/4		3	1		4	16		3	2
	4	1/2		4	−1/3		8	64		4	1

ANSWER:

(a). In (a) each term in this table is found by multiplying the previous term by 2. In fact the entries in this table have the form $(1/16)2^{x-1}$. Answer (b) is incorrect since each term in this table is found by multiplying the previous term by $-1/3$, but exponential functions need ab^x, where $b > 0$. Although in (c) the ratio of adjacent terms is constant, the x values do not change by a constant amount. Answer (d) is incorrect since the ratio between adjacent terms is not constant.

COMMENT:

Have your students solve for a and b in $h(x) = ab^x$ in choice (c) using the first two values and then check what happens to the last two values.

11. Let $f(x) = ab^x$, $b > 0$. Then $\dfrac{f(x+h)}{f(x)} =$

   (a) b^h
   (b) h
   (c) $b^{x+h} - b^x$
   (d) a

ANSWER:

(a), since $f(x+h) = ab^{x+h} = ab^x b^h = f(x)b^h$.

COMMENT:

This fact, in a different format, is used to find the derivative of b^x. This introduces the algebraic manipulations required for the definition of the derivative later in the text.

12. How many asymptotes does the function $f(x) = 1 - 10^{-x}$ have?

   (a) Zero
   (b) One
   (c) Two
   (d) Three

ANSWER:

(b). The only asymptote is a horizontal one, namely $y = 1$.

COMMENT:

Show that $f(x)$ may be thought of as 10^{-x} reflected across the x-axis and then shifted up 1 unit.

13. Estimate the half-life for the exponential decay shown in Figure 1.8.

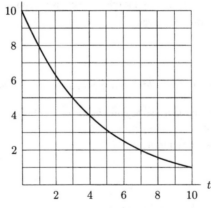

Figure 1.8

ANSWER:

3

COMMENT:

Depending on how much you enlarge the graph, some variation in answers should be allowed here.

14. Estimate the doubling time for the exponential growth shown in Figure 1.9.

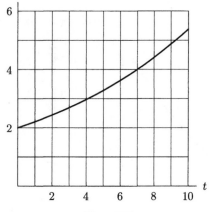

Figure 1.9

ANSWER:

7

COMMENT:

Depending on how much you enlarge the graph, some variation in answers should be allowed here.

15. List at least 5 properties of exponential functions of the form $y(x) = ab^x$ for $a > 0$ and either $0 < b < 1$ or $b > 1$.

ANSWER:

(a) Domain: all real numbers

(b) Range: $y > 0$

(c) Horizontal asymptote: $y = 0$

(d) $y(0) = a$

(e) Always concave up

(f) Always increasing for $b > 1$, and always decreasing for $0 < b < 1$

(g) No vertical asymptote

COMMENT:

You can have your students list more properties. Discuss what happens if $a < 0$.

16. Which of the following graphs is that of $y = ab^x$ if $b > 1$?

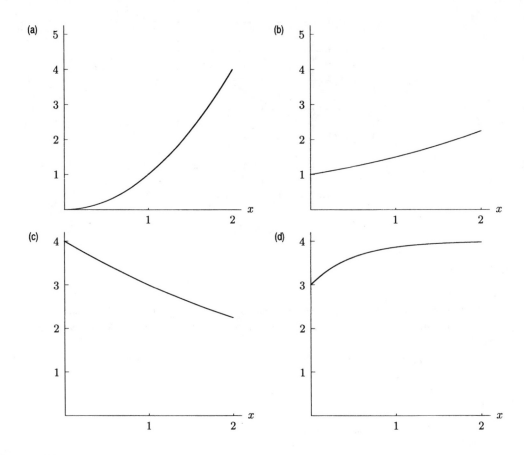

ANSWER:

(b). If $b > 1$, we need a graph that is increasing, concave up, and does not go through the origin.

COMMENT:

Here it is worth having the students point out specifically why the other choices do not work.

Follow-up Questions. What is the value of a? How does the graph change if $a < 0$?

17. *"During 1988, Nicaragua's inflation rate averaged 1.3% a day."* Which formula represents the above statement? Assume t is measured in days.

(a) $I = I_0 e^{0.013t}$

(b) $I = I_0 (1.013)^t$

(c) $I = I_0 (1.013)t$

(d) $I = I_0 (1.3)^t$

ANSWER:

(b)

COMMENT:

Follow-up Question. What happens if the statement is changed to *"During 1988, Nicaragua's inflation rate grew continuously at a rate of 1.3% each day."*?

ConcepTests and Answers and Comments for Section 1.3 ━━━━━━━━

For Problems 1–3, the graph in Figure 1.10 is that of $y = f(x)$. Also, use the graphs (I)–(IV) for the answers.

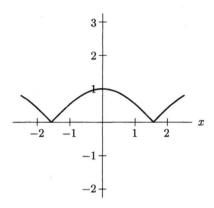

Figure 1.10

(I)
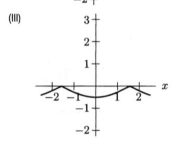

(II)
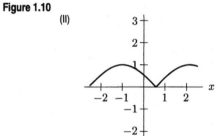

(III)

(IV)
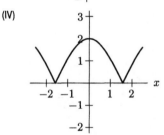

1. Which could be a graph of $cf(x)$?
    ANSWER:
    (III) and (IV). In (III) the graph could be $-1/2 f(x)$ and in (IV) the graph could be $2f(x)$.
    COMMENT:
    You could ask students to verbalize the relationship between $f(x)$ and $cf(x)$ for $c > 1, 0 < c < 1$, and $c < -1$.

2. Which could be a graph of $f(x) - k$?
    ANSWER:
    (I) could be $f(x) - 1$.
    COMMENT:
    You could ask students to verbalize the relationship between $f(x)$ and $f(x) - k$ for $k > 0$ and $k < 0$.

3. Which could be a graph of $f(x - h)$?
    ANSWER:
    (II) could be a graph of $f(x + 1)$.
    COMMENT:
    You could ask students to verbalize the relationship between $f(x)$ and $f(x - h)$ for $h > 0$ and $h < 0$.

4. Which of the following functions is the sum of the functions in Figures 1.11 and 1.12?

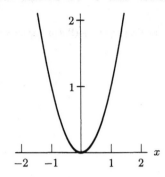

Figure 1.11

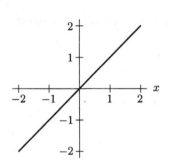

Figure 1.12

(a)

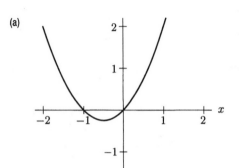

(b)

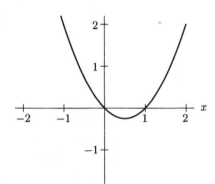

(c)

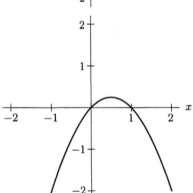

(d)

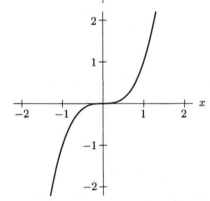

ANSWER:
(a). On the interval $-1 < x < 0$, the sum should be negative and concave up.
COMMENT:
You could follow up by asking for the graph of the difference of the functions in Figures 1.11 and 1.12 (both differences).

5. Given the graph of $y = \sin x$ in Figure 1.13, determine which of the graphs are those of $\sin(2x)$ and $\sin(3x)$?

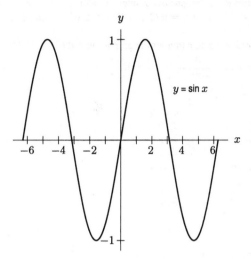

$y = \sin x$

Figure 1.13

(I)

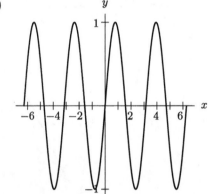

(II)

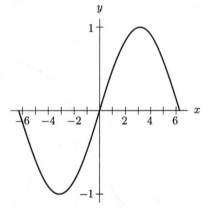

(III)

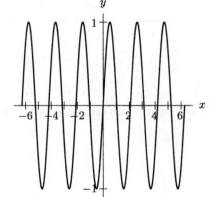

(IV)

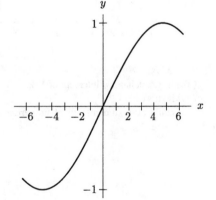

(a) (I) = $\sin(2x)$ and (II) = $\sin(3x)$
(b) (I) = $\sin(2x)$ and (III) = $\sin(3x)$
(c) (II) = $\sin(2x)$ and (III) = $\sin(3x)$
(d) (II) = $\sin(2x)$ and (IV) = $\sin(3x)$
(e) (III) = $\sin(2x)$ and (IV) = $\sin(3x)$

ANSWER:

(b). Replacing x by $2x$ means the first positive x-intercept will be at $x = \pi/2$, so (I) is that of $y = \sin(2x)$. Similarly the first positive zero of $y = \sin(3x)$ is at $x = \pi/3$, so (III) is that of $y = \sin(3x)$.

COMMENT:

Have students also determine the equation for the graphs labeled (II) and (IV).

6. Which of the graphs is that of $y = \dfrac{3}{2 + 4e^{-x}}$?

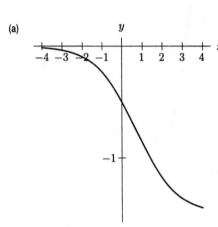
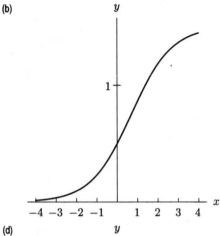
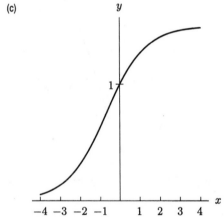
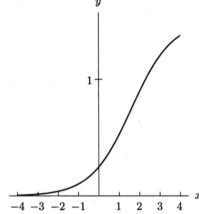

ANSWER:

(b) The graph has a y-intercept of $1/2$.

COMMENT:

You could have students tell of properties evident in the graphs of the other choices that conflict with those of $y = 3/(2 + 4e^{-x})$.

For Problems 7–13, let f and g have values given in the table.

x	$f(x)$	$g(x)$
-2	1	-1
-1	0	1
0	-2	2
1	2	0
2	-1	-2

7. $f(g(1)) =$

   ANSWER:

   $g(1) = 0$, so $f(g(1)) = f(0) = -2$.

   COMMENT:

   You can also consider $f(g(x))$ for $x = -2$ and $x = 2$.

8. $f(g(0)) =$

   ANSWER:

   $g(0) = 2$, so $f(g(0)) = f(2) = -1$.

   COMMENT:

   You can also consider $f(g(x))$ for $x = -2$ and $x = 2$.

9. $f(g(-1)) =$

   ANSWER:

   $g(-1) = 1$, so $f(g(-1)) = f(1) = 2$.

   COMMENT:

   You can also consider $f(g(x))$ for $x = -2$ and $x = 2$.

10. If $f(g(x)) = 1$, then $x =$

    ANSWER:

    $f(-2) = 1$, and $g(2) = -2$, so $x = 2$.

    COMMENT:

    You can also consider $f(g(x)) = a$ for $a = -2, -1, 2$.

11. If $f(g(x)) = 0$, then $x =$

    ANSWER:

    $f(-1) = 0$, and $g(-2) = -1$, so $x = -2$.

    COMMENT:

    You can also consider $f(g(x)) = a$ for $a = -2, -1, 2$.

12. If $g(f(x)) = 2$, then $x =$

    ANSWER:

    $g(0) = 2$, and $f(-1) = 0$, so $x = -1$.

    COMMENT:

    You can also consider $g(f(x)) = a$ for $a = -1, 0, 1$.

13. If $g(f(x)) = -2$, then $x =$

    ANSWER:

    $g(2) = -2$, and $f(1) = 2$, so $x = 1$.

    COMMENT:

    You can also consider $g(f(x)) = a$ for $a = -1, 0, 1$.

For Problems 14–18, let the graphs of f and g be as shown in Figure 1.14. Estimate the values of the following composite functions to the nearest integer.

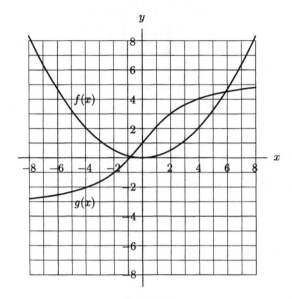

Figure 1.14

14. $g(f(0)) \approx$

    ANSWER:

    $f(0) \approx 0$, so $g(f(0)) \approx g(0) \approx 1$.

    COMMENT:

    You may want to point out that since $f(0) = 0$, this composition is similar to composing a function with the identity function.

15. $g(f(8)) \approx$

    ANSWER:

    $f(8) \approx 8$, so $g(f(8)) \approx g(8) \approx 5$.

    COMMENT:

    You may want to point out that since $f(8) = 8$, this composition is similar to composing a function with the identity function.

16. $g(f(3)) \approx$

    ANSWER:

    $f(3) \approx 1$, so $g(f(3)) \approx g(1) \approx 2$.

    COMMENT:

    When you are computing $g(f(a))$ from the graphs of g and f, it is not always necessary to compute $f(a)$. For example, when the horizontal and vertical scales are the same, you can measure the height of $f(a)$ with a straightedge. This distance placed on the x-axis is the new value from which to measure the height of g. The result will be $g(f(a))$.

17. $f(g(2)) \approx$

    ANSWER:

    $g(2) \approx 3$, so $f(g(2)) \approx f(3) \approx 1$.

    COMMENT:

    See the Comment for Problem 16.

18. $f(g(-1)) \approx$

    ANSWER:

    $g(-1) \approx 0$, so $f(g(-1)) \approx f(0) \approx 0$.

    COMMENT:

    See the Comment for Problem 16.

19. If $f(x) = \sqrt{x^2 + 1}$ and $g(x) = e^{x^2}$ then $f(g(x)) =$

(a) $e^{(x^2+1)}$

(b) $\sqrt{e^{2x^2} + 1}$

(c) $e^{\sqrt{x^2+1}}$

(d) $\sqrt{e^{x^4} + 1}$

ANSWER:

(b)

COMMENT:

Students have trouble simplifying $\left(e^{x^2}\right)^2$. Next you could have them compute $g(f(x))$.

20. For which values of m, n, and b is $f(g(x)) = g(f(x))$ if $f(x) = x + n$ and $g(x) = mx + b$?

(a) $m = 1$, n and b could be any number

(b) $n = 1$, m and b could be any number

(c) $n = 0$, m and b could be any number

(d) $m = 1$, n and b could be any number, or $n = 0$, m and b could be any number.

ANSWER:

(d). After the composition we have: $mx + b + n = mx + mn + b$. Thus $mn = n$ which implies $m = 1$ or $n = 0$.

COMMENT:

This provides an opportunity to point out the difference between subtracting the same quantity from both sides (always allowable) and canceling the same term from both sides (sometimes letting you lose information like $n = 0$).

21. Given the graphs of the functions g and f in Figures 1.15 and 1.16, which of the following is a graph of $f(g(x))$?

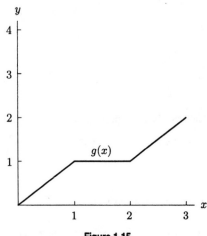

Figure 1.15

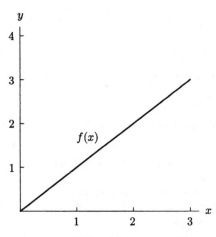

Figure 1.16

(a)

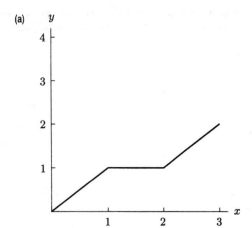

(b)

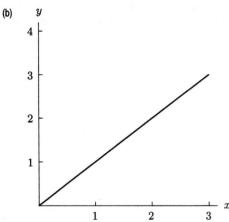

(c)

(d)

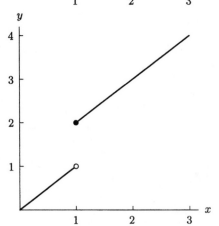

ANSWER:

(a). Because $f(x) = x$, we have $f(g(x)) = g(x)$.

COMMENT:

Follow-up Question. Which graph represents $g(f(x))$?

For Problems 22–23, consider the four graphs.

(I)

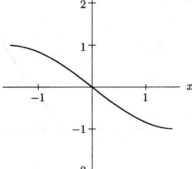

(II)

(III)

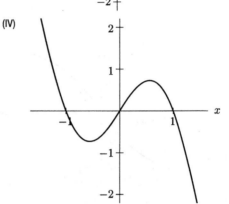

(IV)

22. Which of these graphs could represent even functions?
     ANSWER:
     (III) could be the graph of an even function.
     COMMENT:
     You could ask students to give geometric and analytic definitions here.

23. Which of these graphs could represent odd functions?
     ANSWER:
     (II) and (IV) could be graphs of odd functions.
     COMMENT:
     You could ask students to give geometric and analytic definitions here.

24. Which of the following could be graphs of functions that have inverses?

(a)
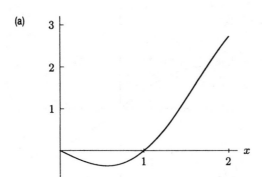

(b)
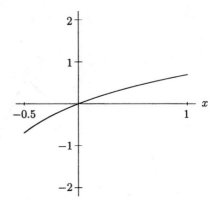

(c)
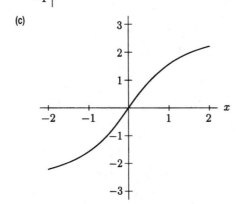

(d)
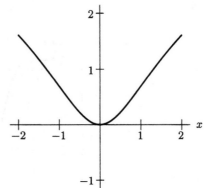

ANSWER:

(b) and (c). Because these graphs pass the horizontal line test, they could have inverses.

COMMENT:

You might redefine the other two functions by limiting their domains so they will have inverses.

25. If $P = f(t) = 3 + 4t$, find $f^{-1}(P)$

ANSWER:

If $P = 3 + 4t$, then $P - 3 = 4t$ and $t = \dfrac{P - 3}{4}$. Thus $f^{-1}(P) = \dfrac{P - 3}{4}$.

COMMENT:

You could also use $f(t) = 3 + 8t^3$, which uses a bit more algebra.

26. Which of the following graphs represents the inverse of the function graphed in Figure 1.17?

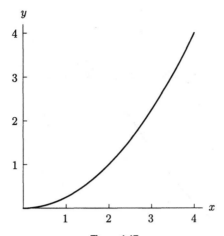

Figure 1.17

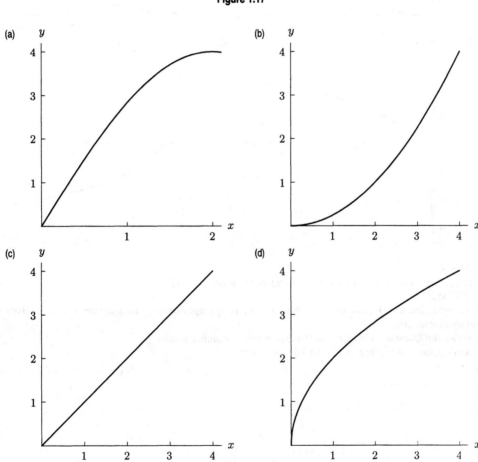

ANSWER:

(d). The graph of the inverse is a reflection of the function across the line $y = x$.

COMMENT:

You could ask the students why each of the other choices fails to be the inverse.

ConcepTests and Answers and Comments for Section 1.4 ━━━━━━

1. Which is a graph of $y = \ln x$?

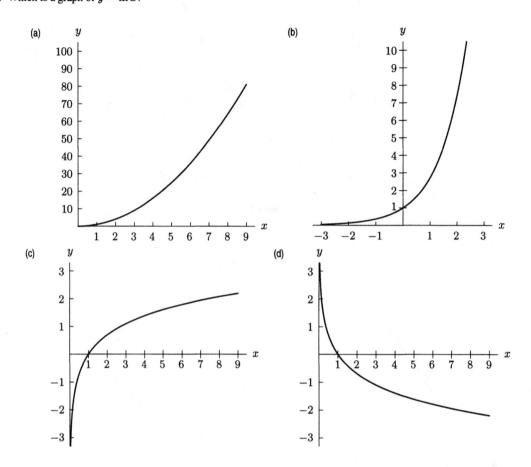

ANSWER:

(c). $y = \ln x$ is an increasing function passing through the point $(1, 0)$.

COMMENT:

You could discuss which properties each of the remaining graphs possess that made the student conclude that it was not an appropriate choice.

Follow-up Question. Find possible formulas for the remaining graphs.

Answer. (a) $y = x^2$, (b) $y = e^x$, and (d) $y = -\ln x$.

2. The graph in Figure 1.18 could be that of

(a) $y = \ln x + \dfrac{1}{2}$

(b) $y = \ln x - \dfrac{1}{2}$

(c) $y = \ln\left(x + \dfrac{1}{2}\right)$

(d) $y = \ln\left(x - \dfrac{1}{2}\right)$

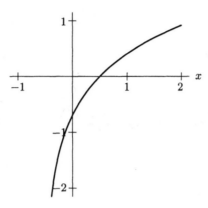

Figure 1.18

ANSWER:

(c). Note that (a) is the graph of $\ln x$ shifted up $\frac{1}{2}$, (b) is shifted down $\frac{1}{2}$, (d) is shifted to the right $\frac{1}{2}$.

COMMENT:

You could also distinguish the four graphs by their horizontal intercepts.

3. Which of the following functions have vertical asymptotes of $x = 3$?

(a) $y = \ln\left(\dfrac{x}{3}\right)$

(b) $y = \ln(x - 3)$

(c) $y = \ln(x + 3)$

(d) $y = 3\ln x$

ANSWER:

(b). Note that (a) and (d) have vertical asymptotes at $x = 0$, while (c) has one at $x = -3$, and (b) has one at $x = 3$, as desired.

COMMENT:

Follow-up Question. Do any of these functions have horizontal asymptotes? If so, what are they?

Answer. No, the range of these functions is all real numbers.

4. Use the properties of logarithms to decide which of the following is largest.

(a) $\ln(30) - \ln(2)$

(b) $2\ln 4$

(c) $\ln 3 + \ln 4$

(d) $\dfrac{\ln 4}{\ln 2}$

ANSWER:

(b). $\ln(30) - \ln(2) = \ln(15)$, $2\ln(4) = \ln(16)$, $\ln(3) + \ln(4) = \ln(12)$, and $\frac{\ln(4)}{\ln(2)} = \frac{\ln(2^2)}{\ln 2} = \frac{2\ln 2}{\ln 2} = 2 = \ln(e^2)$.

Since $e^2 < 9$ and $\ln x$ is an increasing function, $\ln(16)$ is the largest number.

COMMENT:

If they use technology on this problem, then you can point out that using logarithm properties allows exact comparison.

5. The graph of a logarithmic function has a horizontal asymptote.

   (a) True
   (b) False

   ANSWER:

   (b). The range of logarithmic functions consists of all real numbers.

   COMMENT:

   You could also ask about vertical asymptotes.

6. $\log\left(\dfrac{M-N}{M+N}\right) =$

   (a) $2\log M$
   (b) $2\log N$
   (c) $-2\log N$
   (d) $\log(M-N) - \log(M+N)$

   ANSWER:

   (d)

   COMMENT:

   Follow-up Question. What is the value of $\log\left(\dfrac{M^2 - N^2}{M - N}\right)$?

   Answer. $\log(M+N)$.

7. If $\log_{10}(x - a) = n$, then $x =$

   (a) 10^{a+n}
   (b) $a + 10^n$
   (c) $n + 10^a$
   (d) $n + a^{10}$

   ANSWER:

   (b). Compose each side with the exponential function 10^x since it is the inverse function of $\log_{10} x$.

   COMMENT:

   You could ask the same question with the natural logarithm rather than the logarithm base 10.

8. Which of the following functions are increasing and concave up?

   (a) 3^{-x}
   (b) 3^x
   (c) $\ln x$
   (d) $-\ln x$

   ANSWER:

   (b). Note that (a) and (d) are decreasing, and (c) is concave down.

   COMMENT:

   These are all examples of monotonic functions. You could also ask about asymptotes (horizontal and vertical) and intercepts for all four functions.

9. Which of the following functions are decreasing and concave up?

   (a) $-\ln(4 + x)$
   (b) 3^{x-4}
   (c) 3^{4-x}
   (d) $\ln(4 - x)$

   ANSWER:

   (a) and (c). Note that (b) is increasing and (d) is concave down.

   COMMENT:

   You could also ask about asymptotes (horizontal and vertical) and intercepts for all four functions.

10. Which of the following does *not* have a horizontal asymptote?

(a) $y = \log x$

(b) $y = \dfrac{1}{x}$

(c) $y = 5^x$

(d) $y = x^{1/3}$

ANSWER:

(a) and (d). The range of $\log x$ and $x^{1/3}$ is all real numbers.

COMMENT:

Follow-up Question. Which of the above functions does *not* have any asymptotes?

Answer. (d). The domain and range of $y = x^{1/3}$ is all real numbers.

11. Give a formula for the inverse of the following function:

$$P = f(t) = 16e^{14t}$$

(a) $f^{-1}(P) = \dfrac{1}{16}e^{-14P}$

(b) $f^{-1}(P) = \left(\dfrac{\ln 16}{14}\right) P$

(c) $f^{-1}(P) = \dfrac{\frac{1}{14}\ln P}{\ln 16}$

(d) $f^{-1}(P) = \dfrac{1}{14}\ln\left(\dfrac{P}{16}\right)$

ANSWER:

(d). If $P = 16e^{14t}$, then $\frac{P}{16} = e^{14t}$ and $\ln\left(\frac{P}{16}\right) = 14t$. This gives $t = \frac{1}{14}\ln\left(\frac{P}{16}\right)$.

COMMENT:

Students may find that $f^{-1}(P) = \dfrac{1}{14}\left(\ln P - \ln 16\right)$. This would be an excellent time to review the properties of logarithms. For an alternate question, you could also use $f(t) = 2 - e^{-3t}$.

12. Give a formula for the inverse of the following function:

$$P = f(t) = 16\ln(14t)$$

(a) $f^{-1}(P) = \dfrac{1}{14}e^{16P}$

(b) $f^{-1}(P) = \dfrac{1}{14}e^{P/16}$

(c) $f^{-1}(P) = \dfrac{1}{14}\ln\left(\dfrac{P}{16}\right)$

(d) $f^{-1}(P) = \left(\dfrac{\ln 16}{14}\right) P$

ANSWER:

(b). If $P = 16\ln(14t)$, then $\frac{P}{16} = \ln(14t)$ and $e^{P/16} = 14t$. This gives $t = \dfrac{e^{P/16}}{14}$.

COMMENT:

You could also use $f(t) = 6 + 2\ln(3t - 1)$.

13. Solve for x if $8y = 3e^x$.

(a) $x = \ln 8 + \ln 3 + \ln y$

(b) $x = \ln 3 - \ln 8 + \ln y$

(c) $x = \ln 8 + \ln y - \ln 3$

(d) $x = \ln 3 - \ln 8 - \ln y$

ANSWER:

(c). If $8y = 3e^x$, $\frac{8y}{3} = e^x$ and $\ln\left(\frac{8y}{3}\right) = x$.

COMMENT:

This is a good place to point out the many ways answers can be expressed using logarithms.

14. Solve for x if $y = e + 2^x$.

    (a) $x = \dfrac{\ln y - 1}{\ln 2}$

    (b) $x = \dfrac{\ln(y - 1)}{\ln 2}$

    (c) $x = \dfrac{\ln y}{\ln 2} - 1$

    (d) $x = \dfrac{\ln(y - e)}{\ln 2}$

    ANSWER:

    (d). If $y = e + 2^x$, $y - e = 2^x$ and $\ln(y - e) = x \ln 2$. This gives $x = \frac{\ln(y-e)}{\ln 2}$.

    COMMENT:

    You could ask what errors could have been made in obtaining the other choices.

15. For what value of x is $3 \cdot 3^{-x} + 4 = 16 - 3^{-x}$?

    ANSWER:

    If $3 \cdot 3^{-x} + 4 = 16 - 3^{-x}$, then $(3 + 1)3^{-x} = 16 - 4 = 12$. Division gives $3^{-x} = 3$, so $-x = 1$ and $x = -1$.

    COMMENT:

    Students may try to take logarithms of both sides of the original equation.

ConcepTests and Answers and Comments for Section 1.5

1. The amplitude and period of the graph of the periodic function in Figure 1.19 are

	Amplitude	Period
(a)	2	2
(b)	2	3
(c)	2	1/2
(d)	3	2
(e)	3	1/2

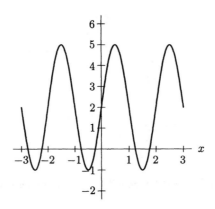

Figure 1.19

    ANSWER:

    (d). The maximum value of the function is 5, and the minimum value is -1, so the amplitude is $\dfrac{5 - (-1)}{2} = 3$. The function repeats itself after 2 units, so the period is 2.

    COMMENT:

    Point out that the function is oscillating about the line $y = 2$. Have the students find a formula for the function shown in Figure 1.19.

2. The amplitude and period of the graph of the periodic function in Figure 1.20 are

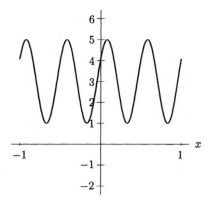

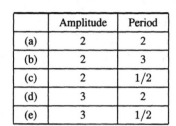

	Amplitude	Period
(a)	2	2
(b)	2	3
(c)	2	1/2
(d)	3	2
(e)	3	1/2

Figure 1.20

ANSWER:

(c). The maximum value of the function is 5 and the minimum is 1, so the amplitude is $\dfrac{5-1}{2} = 2$. The function repeats itself after $1/2$ unit, so the period is $1/2$.

COMMENT:

It is easiest to find the period using the extreme values of the function.

3. Which of the following could describe the graph in Figure 1.21.

(a) $y = 3\sin\left(\dfrac{x}{2} + \dfrac{\pi}{2}\right)$

(b) $y = 3\sin\left(2x + \dfrac{\pi}{2}\right)$

(c) $y = 3\cos(2x)$

(d) $y = 3\cos\left(\dfrac{x}{2}\right)$

(e) $y = 3\sin(2x)$

(f) $y = 3\sin\left(\dfrac{x}{2}\right)$

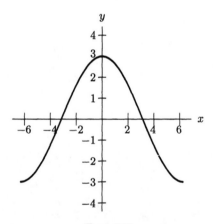

Figure 1.21

ANSWER:

(a) and (d). Note that (b), (c), and (e) have period π, with the rest having period 4π. Answer (f) has $y(0) = 0$. (a) and (d) could describe the graph.

COMMENT:

The fact that the same graph may have more than one analytic representation could be emphasized here.

4. Figure 1.22 shows the graph of which of the following functions?

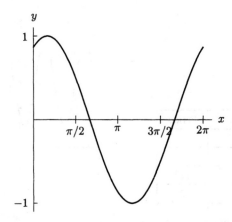

Figure 1.22

(a) $y = \cos(x + \pi/6)$
(b) $y = \cos(x - \pi/6)$
(c) $y = \sin(x - \pi/6)$
(d) $y = \sin(x + \pi/6)$
(e) None of these

ANSWER:
(b)
COMMENT:
You could ask what the graphs of the other choices look like.

5. Which of the graphs are those of $\sin(2t)$ and $\sin(3t)$?

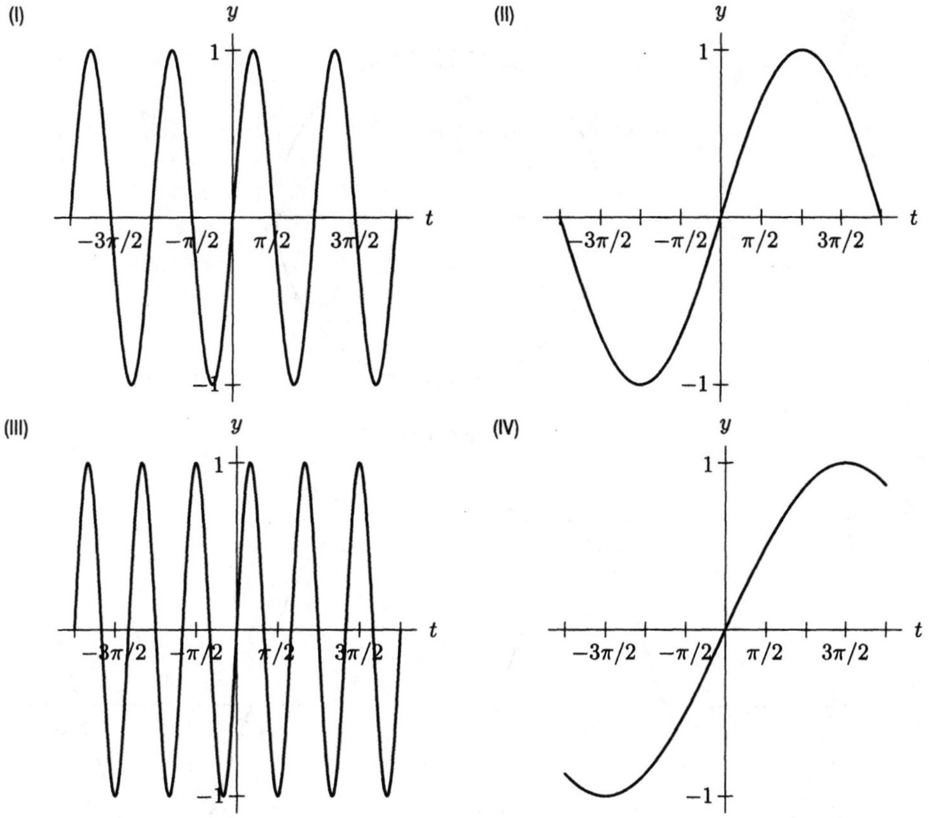

(I)

(II)

(III)

(IV)

(a) (I) = $\sin(2t)$ and (II) = $\sin(3t)$

(b) (I) = $\sin(2t)$ and (III) = $\sin(3t)$

(c) (II) = $\sin(2t)$ and (III) = $\sin(3t)$

(d) (III) = $\sin(2t)$ and (IV) = $\sin(3t)$

ANSWER:

(b). The period is the time needed for the function to execute one complete cycle. For $\sin(2t)$, this will be π and for $\sin(3t)$, this will be $2\pi/3$.

COMMENT:

You could ask what equations describe the graphs of (II) and (IV). This question is the same as Question 5 in Section 1.3.

6. Consider a point on the unit circle (shown in Figure 1.23) starting at an angle of zero and rotating counterclockwise at a constant rate. Which of the graphs represents the horizontal component of this point as a function of time.

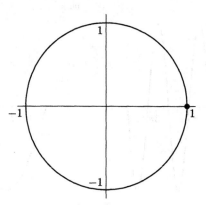

Figure 1.23

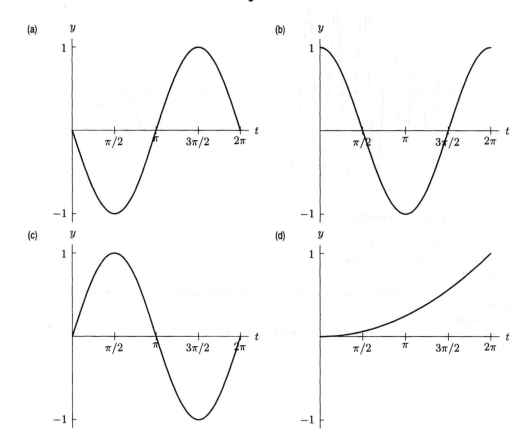

ANSWER:

(b)

COMMENT:

Use this question to make the connection between the unit circle and the graphs of trigonometric functions stronger.

Follow-up Question. Which of the graphs represents that of the vertical component of this point as a function of time?

Answer. (c)

7. Which of the following is the approximate value for the sine and cosine of angles A and B in Figure 1.24?

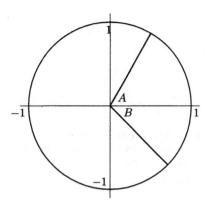

Figure 1.24

(a) $\sin A \approx 0.5$, $\cos A \approx 0.85$, $\sin B \approx -0.7$, $\cos B \approx 0.7$
(b) $\sin A \approx 0.85$, $\cos A \approx 0.5$, $\sin B \approx -0.7$, $\cos B \approx 0.7$
(c) $\sin A \approx 0.5$, $\cos A \approx 0.85$, $\sin B \approx 0.7$, $\cos B \approx 0.7$
(d) $\sin A \approx 0.85$, $\cos A \approx 0.5$, $\sin B \approx 0.7$, $\cos B \approx 0.7$

ANSWER:

(b)

COMMENT:

You could use this question to make the connection between the unit circle and the trigonometric functions stronger. You could also identify other points on the circle and ask for values of sine or cosine.

8. Figure 1.25 shows the graph of which of the following functions?

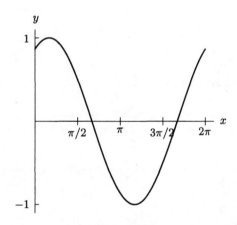

Figure 1.25

(a) $y = \sin(x - \pi/3)$
(b) $y = \cos(x - \pi/3)$
(c) $y = \sin(x + \pi/3)$
(d) $y = \cos(x + \pi/3)$
(e) None of these

ANSWER:

(c)

COMMENT:

You could ask what the graphs of the other choices look like.

9. Which of the following are properties of both $y = \arcsin x$ and $y = \dfrac{x}{\sqrt{1-x^2}}$?

   (a) $y(0) = 0$
   (b) Increasing everywhere
   (c) Concave down for $x < 0$
   (d) Domain: $[-1, 1]$
   (e) Range: $(-\infty, \infty)$

   ANSWER:
   (a), (b), and (c). The domain of $y = \dfrac{x}{\sqrt{1-x^2}}$ is $(-1, 1)$ and the range of $\arcsin x$ is $[-\pi/2, \pi/2]$.

   COMMENT:
   You may want to have the students sketch graphs of both functions.
   Follow-up Question. What is the concavity for each function when $x > 0$?
   Answer. Both functions are concave up when $x > 0$.

10. Which of the following have the same domain?

    (a) $\arcsin x$
    (b) $\arctan x$
    (c) $\ln x$
    (d) e^x
    (e) $\sqrt{1 - x^2}$

    ANSWER:
    (a) and (e), also (b) and (d). Note that the domain of $\arcsin x$ and $\sqrt{1 - x^2}$ is $[-1, 1]$, the domain of $\arctan x$ and e^x is $(-\infty, \infty)$.

    COMMENT:
    Students could be thinking of the range of $\sin x$ for (a) and the range of $\tan x$ for (b).
    Follow-up Question. Does changing (c) to $\ln |x|$ affect the answer?
    Answer. No, because the domain of $\ln |x|$ is all real numbers except $x = 0$.

11. Which of the following have the same range?

    (a) $\arcsin x$
    (b) $\arctan x$
    (c) $\ln x$
    (d) e^x
    (e) $\sqrt{1 - x^2}$

    ANSWER:
    (a) and (b). The range of (a) and (b) is $\left[-\dfrac{\pi}{2}, \dfrac{\pi}{2} \right]$.

    COMMENT:
    Follow-up Question. Does changing (c) to $|\ln x|$ affect the answer?
    Answer. No, the range of $|\ln x|$ is $[0, \infty)$ while the range of e^x is $(0, \infty)$.

12. If $y = \arcsin x$, then $\cos y =$

    (a) $\sqrt{1 - x^2}$

    (b) $\sqrt{x^2 - 1}$

    (c) $\dfrac{1}{\sqrt{x^2 - 1}}$

    (d) $\dfrac{1}{\sqrt{1 - x^2}}$

    (e) $\dfrac{1}{x}$

    ANSWER:
    (a). If $y = \arcsin x$, then $-\frac{\pi}{2} < y < \frac{\pi}{2}$ giving $\cos y = \sqrt{1 - \sin^2 y} = \sqrt{1 - x^2}$.
    COMMENT:
    You could also solve this problem by drawing a right triangle with hypotenuse of 1 and labeling the side opposite angle y as x.

13. If $y = \arcsin x$, then $\tan y =$

(a) $\dfrac{\sqrt{1 - x^2}}{x}$

(b) $\dfrac{\sqrt{x^2 - 1}}{x}$

(c) $\dfrac{x}{\sqrt{x^2 - 1}}$

(d) $\dfrac{x}{\sqrt{1 - x^2}}$

(e) $\dfrac{1}{x}$

ANSWER:

(d). If $y = \arcsin x$, then $-\frac{\pi}{2} < y < \frac{\pi}{2}$ giving $\tan y = \dfrac{\sin y}{\cos y} = \dfrac{\sin y}{\sqrt{1 - \sin^2 y}} = \dfrac{x}{\sqrt{1 - x^2}}$.

COMMENT:

You could also solve this problem by drawing a right triangle with hypotenuse of 1 and labeling the side opposite angle y as x.

14. If $y = \arctan x$, then $\cos y =$

(a) $\dfrac{\sqrt{1 + x^2}}{x}$

(b) $\dfrac{x}{\sqrt{x^2 + 1}}$

(c) $\dfrac{1}{\sqrt{x^2 + 1}}$

(d) $\sqrt{1 + x^2}$

(e) $\dfrac{1}{x}$

ANSWER:

(c). If $y = \arctan x$, then $-\frac{\pi}{2} < y < \frac{\pi}{2}$. Because $\tan^2 y + 1 = \dfrac{1}{\cos^2 y}$,

$\cos y = \sqrt{\dfrac{1}{\tan^2 y + 1}} = \sqrt{\dfrac{1}{x^2 + 1}}$.

COMMENT:

You could also solve this problem by drawing a right triangle with the side opposite angle y labeled as x and the side adjacent to angle y labeled as 1.

15. If $y = \arctan x$, then $\sin y =$

(a) $\dfrac{\sqrt{1 + x^2}}{x}$

(b) $\dfrac{x}{\sqrt{x^2 + 1}}$

(c) $\dfrac{1}{\sqrt{x^2 + 1}}$

(d) $\sqrt{1 + x^2}$

(e) $\dfrac{1}{x}$

ANSWER:

(b). If $y = \arctan x$, then $-\frac{\pi}{2} < y < \frac{\pi}{2}$ and $\sin y = \tan y \cos y$. Because $\tan^2 y + 1 = \dfrac{1}{\cos^2 y}$, $\cos y = \sqrt{\dfrac{1}{\tan^2 y + 1}}$

and $\sin y = \dfrac{x}{\sqrt{x^2 + 1}}$.

COMMENT:

You could also solve this problem by drawing a right triangle with the side opposite angle y labeled as x and the side adjacent to angle y labeled as 1.

ConcepTests and Answers and Comments for Section 1.6

1. Graph $y = x^2$, $y = x^3$, $y = x^4$, $y = x^5$. List at *least* 3 observations. *(Closed Book)*

ANSWER:

(a) $y = x^2$ and $y = x^4$ have the same general shape—that of a "U".

(b) $y = x^3$ and $y = x^5$ have the same general shape—that of a "seat".

(c) For $x > 1$, as the power increases, the function grows faster.

(d) When $0 < x < 1$, we have $x^2 > x^3 > x^4 > x^5$.

(e) For $x > 0$, all functions are increasing and concave up.

(f) All functions intersect at $(1, 1)$ and $(0, 0)$.

(g) For $x < 0$, the functions $y = x^2$ and $y = x^4$ are decreasing and concave up.

(h) For $x < 0$, the functions $y = x^3$ and $y = x^5$ are increasing and concave down.

COMMENT:

This question could also be used as an exploratory activity.

2. Graph $y = x^{-1}$, $y = x^{-2}$, $y = x^{-3}$, $y = x^{-4}$. List at *least* 3 observations. *(Closed Book)*
   ANSWER:

   (a) $y = x^{-2}$ and $y = x^{-4}$ have the same general shape and they are always positive.
   (b) $y = x^{-1}$ and $y = x^{-3}$ have the same general shape.
   (c) As the power decreases, the function approaches 0 faster as x increases.
   (d) For $x > 0$, they are all concave up.
   (e) They intersect at $(1, 1)$.
   (f) Each has a vertical asymptote at $x = 0$ and a horizontal asymptote at $y = 0$.
   (g) For $x < 0$, the functions $y = x^{-2}$ and $y = x^{-4}$ are increasing and concave up.
   (h) For $x < 0$, the functions $y = x^{-3}$ and $y = x^{-5}$ are decreasing and concave down.

   COMMENT:
   This question could also be used as an exploratory activity.

3. Graph $y = x^{1/2}$, $y = x^{1/3}$, $y = x^{1/4}$, $y = x^{1/5}$. What do you observe about the growth of these functions? *(Closed Book)*
   ANSWER:
   The smaller the power of the exponent, the slower the function grows for $x > 1$.
   COMMENT:
   Your students may observe other properties.

For Problems 4–8, as $x \to \infty$ which function dominates, (a) or (b)? (That is, which function is larger in the long run?)

4. (a) $0.1x^2$
   (b) $10^{10}x$
   ANSWER:
   (a). Power functions with the power greater than one and with a positive coefficient grow faster than linear functions.
   COMMENT:
   You could ask about the behavior as $x \to -\infty$ as well.

5. (a) $0.25\sqrt{x}$
   (b) $25,000x^{-3}$
   ANSWER:
   (a). Note that $0.25\sqrt{x}$ is an increasing function whereas $25,000x^{-3}$ is a decreasing function.
   COMMENT:
   One reason for such a question is to note that global behavior may not be determined by local behavior.

6. (a) $3 - 0.9^x$
   (b) $\log x$
   ANSWER:
   (b). Note that $3 - 0.9^x$ has a horizontal asymptote whereas the range of $\log x$ is all real numbers.
   COMMENT:
   Students should realize that the graph the calculator displays can be misleading.

7. (a) x^3
   (b) 2^x
   ANSWER:
   (b). Exponential growth functions grow faster than power functions.
   COMMENT:
   You could ask about the behavior as $x \to -\infty$ as well.

8. (a) $10(2^x)$
   (b) $72,000x^{12}$
   ANSWER:
   (a). Exponential growth functions grow faster than power functions, no matter how large the coefficient.
   COMMENT:
   One reason for such a question is to note that global behavior may not be determined by local behavior.

9. List the following functions in order from smallest to largest as $x \to \infty$ (that is, as x increases without bound).

   (a) $f(x) = -5x$
   (b) $g(x) = 10^x$
   (c) $h(x) = 0.9^x$
   (d) $k(x) = x^5$
   (e) $l(x) = \pi^x$

   ANSWER:

   (a), (c), (d), (e), (b). Notice that $f(x)$ and $h(x)$ are decreasing functions, with $f(x)$ being negative. Power functions grow slower than exponential growth functions, so $k(x)$ is next. Now order the remaining exponential functions, where functions with larger bases grow faster.

   COMMENT:

   This question was used as an elimination question in a classroom session modeled after "Who Wants to be a Millionaire?", replacing "Millionaire" by "Mathematician".

10. What is the degree of the graph of the polynomial in Figure 1.26?

   (a) 3
   (b) 5
   (c) Either (a) or (b)
   (d) Neither (a) nor (b)
   (e) Any polynomial of degree greater than 2

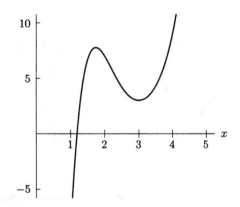

Figure 1.26

   ANSWER:

   (e). The graph could represent any such polynomial. The window may not show all the crucial behavior of the polynomial.

   COMMENT:

   This question is meant to help students realize how little information a graph alone gives. The window may not show all the crucial behavior of the polynomial. You may want to draw other curves and ask for a possible degree if the curve is that of a polynomial.

11. The equation $y = x^3 + 2x^2 - 5x - 6$ is represented by which graph?

(a)
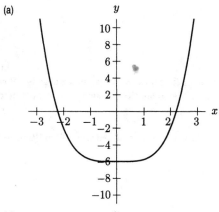

(b)
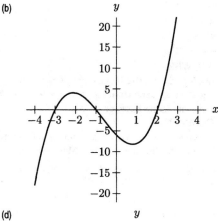

(c)
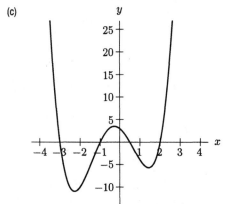

(d)
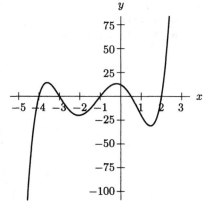

ANSWER:

(b). The graph will have a y-intercept of -6 and not be that of an even function.

COMMENT:

You may want to point out the various tools students can use to solve this problem, i.e. intercepts, even/odd, identifying the zeros, etc. You could also have students identify a property in each of the other choices that is inconsistent with the graph of $y = x^3 + 2x^2 - 5x - 6$.

12. The graph in Figure 1.27 is a representation of which function?

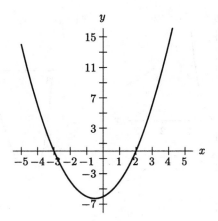

Figure 1.27

(a) $y = 3x + 2$

(b) $y = (x - 2)(x + 3)$

(c) $y = (x - 6)(x - 2)$

(d) $y = (x - 3)(x + 2)$

(e) None of these

ANSWER:

(b). The graph is a parabola with x-intercepts of 2 and -3.

COMMENT:

You could ask students to describe the graphs of the other equations.

13. The equation $y = x^2 + 5x + 6$ is represented by which graph?

(a)

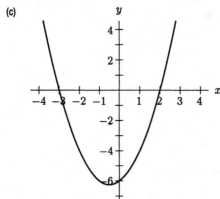

(b)

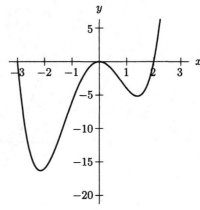

(c)

(d)

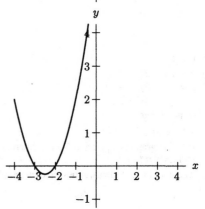

ANSWER:

(d). The graph will be a parabola with x-intercepts of -2 and -3.

COMMENT:

Have students identify a property in each of the other choices that is inconsistent with the graph of $y = x^2 + 5x + 6$.

14. Which of the graphs could represent a graph of $y = ax^4 + bx^3 + cx^2 + dx + e$? Here a, b, c, d and e are real numbers, and $a \neq 0$.

(a)

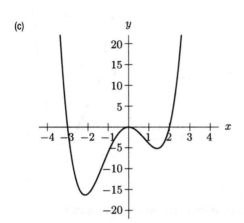

(b)

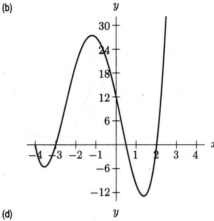

(c)

(d)

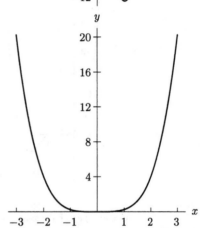

ANSWER:

(a), (b), (c), and (d)

COMMENT:

Follow-up Question. What property could you observe in a graph to know it could not be that of a fourth degree polynomial?

Answer. The graph of the function has opposite end behavior ($y \to \infty$ as $x \to \infty$), or the graph turns more than three times, are a few properties students might observe.

15. Which of the graphs (a)–(d) could be that of a function with a double zero?

(a)
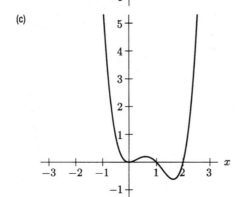

(b)
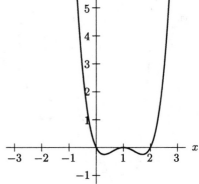

(c)

(d)
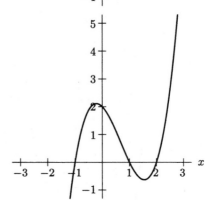

ANSWER:

The graphs in (b) and (c) touch the x-axis, but do not cross there. So, they have double zeros.

COMMENT:

Students should remember that a double zero means the function will "bounce" off the x-axis.

16. Let $f(x) = \dfrac{x^2 - 1}{x + 1}$ and $g(x) = x - 1$, then $f(x) = g(x)$.

(a) True
(b) False

ANSWER:

(b). The domain of $f(x)$ is not equal to the domain of $g(x)$.

COMMENT:

Students should realize that some algebraic manipulations can only be applied to functions if the domain is stated.

Follow-up Question. How can you change the statement to make it true?

Answer. Keep $f(x)$ defined as in the problem, and remove $x = -1$ from the domain of $g(x)$. Alternatively, replace "$f(x) = g(x)$" with "$f(x) = g(x)$ if $x \neq -1$".

17. Without using your calculator, which of the following is a graph for $y = \dfrac{1 - x^2}{x - 2}$?

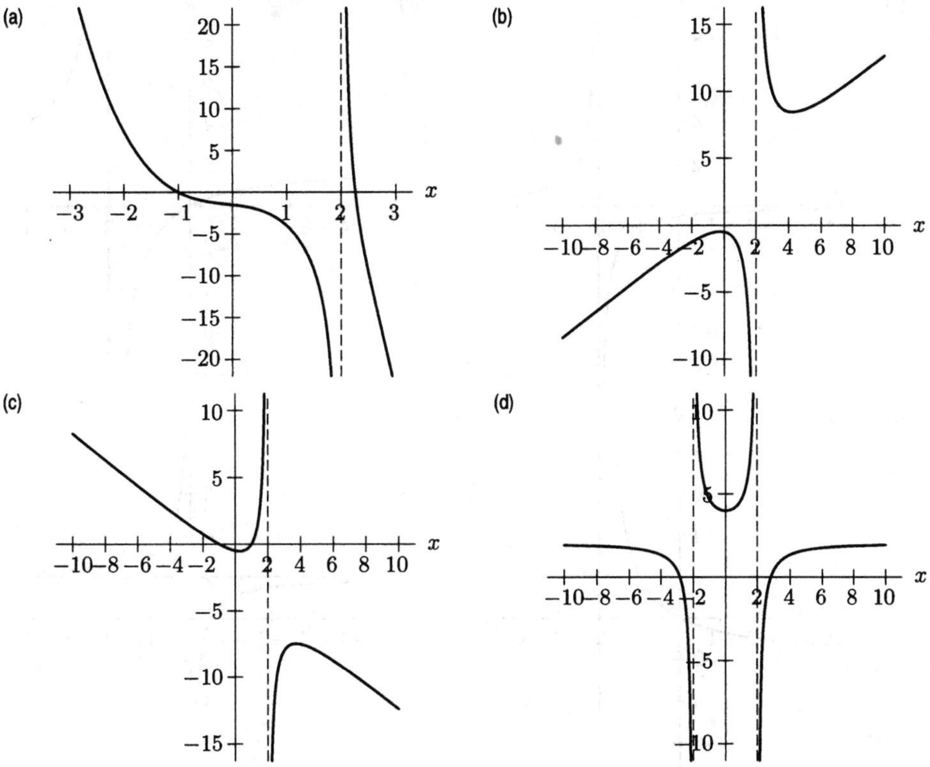

(a)

(b)
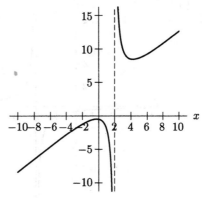

(c)
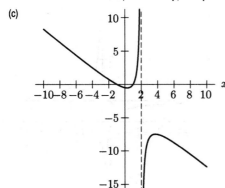

(c)
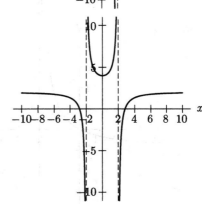

(d)

ANSWER:

(c). The equation indicates x-intercepts at ± 1 and a vertical asymptote at $x = 2$. (c) is the only graph where this happens.

COMMENT:

Students should analyze the long-term behavior of the function as well as the short-term behavior (i.e. asymptotes). You could also ask the students to analyze the similarities and differences of the functions (i.e. zeros, intercepts, etc.).

18. Which of the graphs represents $y = \dfrac{2x}{x-2}$?

(a)

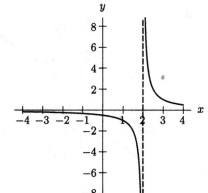

(b)

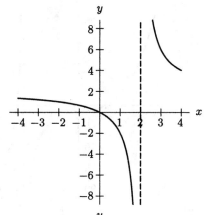

(c)

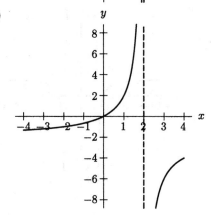

(d)

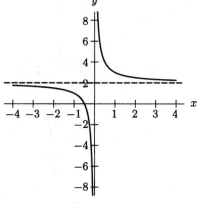

ANSWER:

(b). The graph goes through the origin and is positive for $x > 2$.

COMMENT:

Have students identify a property in each of the other choices that is inconsistent with the graph of $y = \dfrac{2x}{x-2}$.

19. Which of the graphs represents $y = \dfrac{-x}{x^2 + x - 2}$?

(a)

(b)

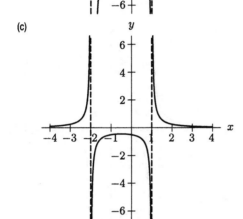

(c)

(d)

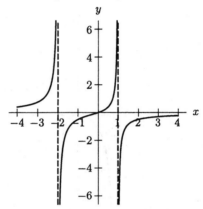

ANSWER:

(d). The graph goes through the origin and has vertical asymptotes at $x = 1$ and -2. It is also positive for $0 < x < 1$, and for large x, the function is negative.

COMMENT:

You could have the students verbalize why they excluded choices (a), (b), and (c).

20. Which of the graphs represents $y = \dfrac{2x^2}{x^2 + x - 2}$?

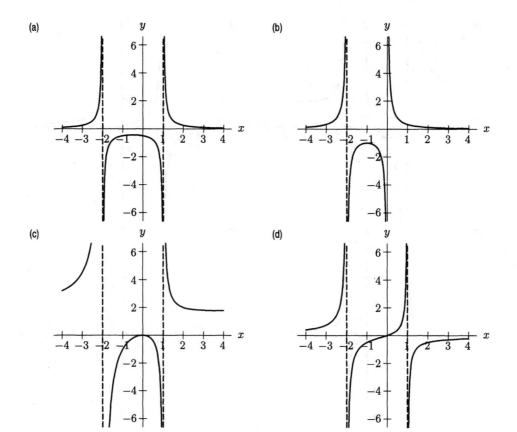

ANSWER:

(c). The graph goes through the origin and has vertical asymptotes at $x = 1$ and -2. It is negative for $0 < x < 1$, and the function is positive for large x. Alternatively, the function has a horizontal asymptote $y = 2$.

COMMENT:

You could have the students verbalize why they excluded choices (a), (b), and (d).

21. Which of the following functions represents the higher of the two functions in Figure 1.28 divided by the lower of the two functions?

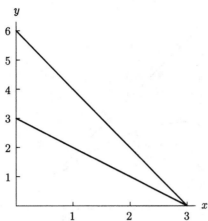

Figure 1.28

(a)
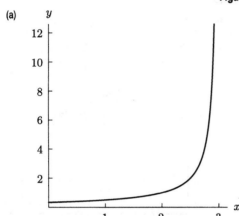

(b)
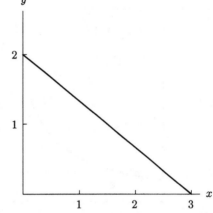

(c)
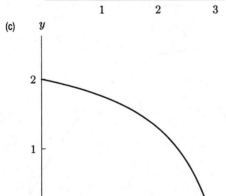

(d)
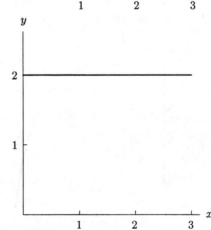

ANSWER:

(d). The ratio of the higher of the two functions to the lower equals 2 at $x = 0$, $x = 1$, and $x = 2$.

COMMENT:

You might want to point out that the division is undefined when $x = 3$.

Follow-up Question. What does the graph of the reciprocal of this ratio look like?

Answer. $y = \begin{cases} \frac{1}{2} & \text{if} \quad x \neq 3 \\ \text{undefined} & \text{if} \quad x = 3. \end{cases}$

22. Which of the following functions represents the higher of the two functions in Figure 1.29 divided by the lower of the two functions?

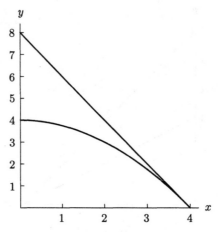

Figure 1.29

(a)

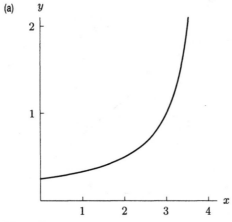

(b)

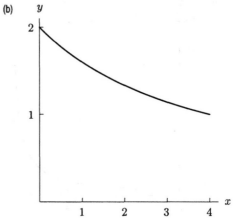

(c)

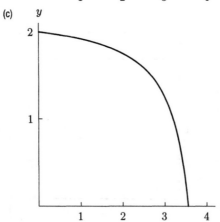

(d)

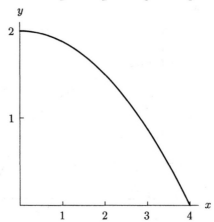

ANSWER:

(b). Notice this ratio starts at 2 and approaches 1 as x approaches 4.

COMMENT:

The division of the two functions is not defined when $x = 4$.

Follow-up Question. Could the graph of the reciprocal of this ratio be one of these choices?

Answer. No, the graph of the reciprocal passes through the point $(0, 1/2)$.

23. Which of the following functions represents the higher of the two functions in Figure 1.30 divided by the lower of the two functions?

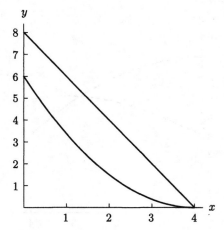

Figure 1.30

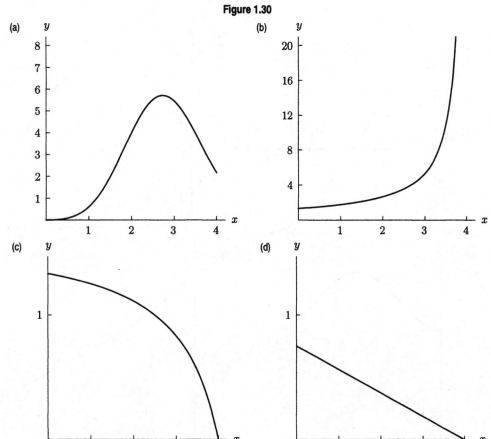

ANSWER:

(b). Because as x increases the lower graph is approaching zero faster than the upper graph, their ratio will be increasing.

COMMENT:

The division of the two functions is not defined when $x = 4$.

Follow-up Question. Could the graph of the reciprocal of this ratio be one of these choices?

Answer. Yes, (d). because the lower function could be a quadratic, and a quadratic function divided by a linear function is linear where it is defined.

24. Which of the following functions represents the ratio of the function starting at the point $(0, 8)$ to the function starting at the point $(0, 4)$ as shown in Figure 1.31.

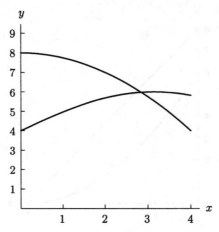

Figure 1.31

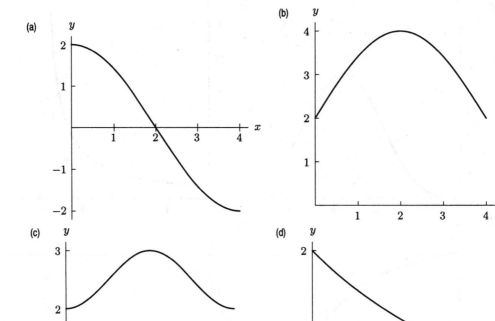

ANSWER:

(d). The ratio is always positive, decreasing, and less than 1 after the curves cross. Alternatively, the ratio has the value 1 when $x \approx 3$.

COMMENT:

Follow-up Question. Could the graph of the reciprocal of this ratio be one of the choices?

Answer. No. none of these graphs passes through the point $(0, 1/2)$.

25. Every exponential function has a horizontal asymptote.

    (a) True
    (b) False

    ANSWER:

    (a). The horizontal axis is a horizontal asymptote for all exponential functions.

    COMMENT:

    You may need to remind students that an exponential function has the form $y = ab^x$ where $a \neq 0$ and $b > 0$ but $b \neq 1$.

26. Every exponential function has a vertical asymptote.

    (a) True
    (b) False

    ANSWER:

    (b). Exponential functions do not have vertical asymptotes. The domain for an exponential function is all real numbers.

    COMMENT:

    You may point out that analyzing the domain of a function is another way to decide if vertical asymptotes exist.

ConcepTests and Answers and Comments for Section 1.7

For Problems 1–6, decide whether the function is continuous on the given interval.

1.
$$f(x) = \frac{1}{x - 2} \qquad \text{on } [0, 3]$$

    ANSWER:

    $f(x)$ is not continuous because $f(2)$ is not defined.

    COMMENT:

    Students should realize that changing the domain affects the answer. You may want to ask for the definition of continuity on an interval.

2.
$$f(x) = \frac{1}{x - 2} \qquad \text{on } [-1, 0]$$

    ANSWER:

    $f(x)$ is continuous.

    COMMENT:

    You may want to ask for the definition of continuity on an interval.

3.
$$f(x) = e^{\tan \theta} \qquad \text{on } \left[-\frac{\pi}{2}, \frac{\pi}{2}\right]$$

    ANSWER:

    $f(x)$ is not continuous since $\tan \theta$ is a not defined for $x = -\frac{\pi}{2}, \frac{\pi}{2}$.

    COMMENT:

    Follow-up Question. Is $f(x) = e^{\tan \theta}$ continuous on $(-\pi/2, \pi/2)$?

4.
$$f(x) = \frac{x^2 - 1}{x + 1} \qquad \text{on } (-\infty, \infty)$$

    ANSWER:

    $f(x)$ is not continuous since $f(-1)$ is not defined.

    COMMENT:

    Ask the same question, but change the interval so that $x = -1$ is not in the interval.

5.

$$f(x) = \begin{cases} \frac{x^2-1}{x+1} & , & x \neq -1 \\ -2 & , & x = -1 \end{cases}$$

Answer $f(x)$ is continuous.
   COMMENT:
   You might spend more time discussing piecewise defined functions.

6. The cost of a first class stamp over the past 100 years.
    ANSWER:
    This function is not continuous. Stamp prices increase by a discrete amount.
    COMMENT:
    Have your students describe other functions that are not continuous.

7. The range of which of the following functions can take on all values between -2 and 2.

   (a) $y = \arcsin(2x)$
   (b) $y = 2\arctan x$
   (c) $y = 1 + \cos x$
   (d) $y = \dfrac{1}{(1-x)^2}$
   (e) $y = \dfrac{1}{9-x^2}$

   ANSWER:

   (b). The range of $y = 2\arctan x$ is $[-\pi, \pi]$, whereas that of $\arcsin(2x)$ is $\left[-\dfrac{\pi}{2}, \dfrac{\pi}{2}\right]$, so the function in (a) cannot
have the value of 2. The functions in (c) and (d) are always positive, while (e) can never equal 0.
   COMMENT:
   Follow-up Question. Which functions have the same range?
   Answer. The range of (b) and (c) is all real numbers.

ConcepTests and Answers and Comments for Section 1.8 ——————————————

1. Possible criteria for a limit: *As you get closer and closer to the limit point the function gets closer and closer to the limit
value.* Which of the following is an example that meets the criteria but does not have the stated limit?

   (a) As x increases to 2, $f(x) = x^2$ gets closer and closer to 4, so the limit at $x = 2$ of $f(x)$ is 4.
   (b) As x increases to 100, $f(x) = 1/x$ gets closer and closer to 0, so the limit as x goes to 100 of $f(x)$ is 0.
   (c) As x increases to 3, $f(x) = (1 + x)^2$ gets closer and closer to 16, so the limit as x goes to 3 of $f(x)$ is 16.
   (d) None of these show a problem with this criteria for a limit.

   ANSWER:
   (b) meets this criteria, but this limit is $1/100$, not 0.
   COMMENT:
   If you do not want to talk about one-sided limits, then you may not want to use this question.

2. Which of the graphs has an ϵ and δ which would satisfy the definition of a limit at $x = c$?

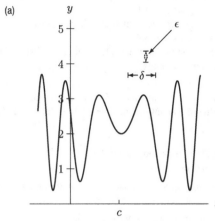

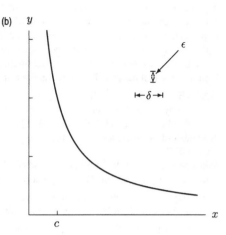

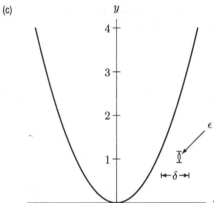

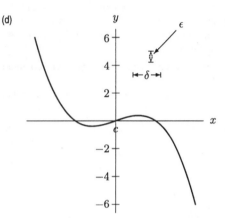

ANSWER:

(d)

COMMENT:

You might want to make sure that students realize that the δ and ϵ shown should be placed on both sides of c and $f(c)$, respectively. You can give students a specific $y = f(x)$ to enter in their calculator, along with a range of $f(c) - \epsilon < y < f(c) + \epsilon$ and ask them to find the x-window so the graph of f exits the side (rather than the top). Then ask them to explain how that could give them a value δ which works in the limit definition.

3. Give an example of a function showing why each of the following could not be the only requirement in the definition of a limit.

(a) $\lim\limits_{x \to a} f(x) = L$ if $f(a) = L$

(b) $\lim\limits_{x \to a} f(x) = L$ if $f(x)$ becomes closer to L as x becomes closer to a

ANSWER:

(a) Consider $f(x) = \begin{cases} 1/(x - a) & , \quad x \neq a \\ L & , \quad x = a \end{cases}$

(b) Consider $f(x) = (x - a)^2$ and $L = -1$

COMMENT:

You could have students describe these examples graphically instead of giving equations.

4. Which of the following statements is true of the limit of the function $\sin(1/x)$ as x goes to zero?

   (a) If you give me a number, ϵ, no matter how small, I can always get close enough to $x = 0$ so that $\sin(1/x)$ stays within that number of 0. Therefore the limit of $\sin(1/x)$ as x goes to 0 is 0.

   (b) If you give me a number, ϵ, no matter how small, I can always get close enough to $x = 0$ so that $\sin(1/x)$ stays within that number of 1. Therefore the limit of $\sin(1/x)$ as x goes to 0 is 1.

   (c) If you give me a number, ϵ, no matter how small, there is no point on the x-axis close enough to 0 such that $\sin(1/x)$ is confined to that number. Therefore the limit of $\sin(1/x)$ does not exist.

   ANSWER:

   (c). For any interval containing 0, the value of $\sin(1/x)$ takes on the value of 0 an infinite number of times (at $x = 1/(n\pi)$, where n is an integer) and $\sin(1/x)$ takes the value of 1 an infinite number of times (at $x = 2/(m\pi)$ where m is a member of the sequence $1, 5, 9, 13, \ldots$).

   COMMENT:

   Follow-up Question. What happens to the graph of $x \sin(1/x)$ as x approaches 0?

   Answer. The graph of $x \sin(1/x)$ approaches 0 as x approaches 0.

5. Possible criteria for continuity of a function at a point: *If the limit of the function exists at a point, the function is continuous at that point.* Which of the following examples fits the above criteria but is not continuous at $x = 0$?

   (a) $f(x) = x$. The limit of $f(x)$, as x goes to 0, is 0 so this function is continuous at $x = 0$.

   (b) $f(x) = x^2/x$. The limit of $f(x)$, as x goes to 0, is 0 therefore $f(x)$ is continuous at $x = 0$.

   (c) $f(x) = |x|/x$. The limit of $f(x)$, as x goes to 0, is 1 therefore $f(x)$ is continuous at $x = 0$.

   (d) None of these show a problem with this criteria.

   ANSWER:

   (b). $f(0)$ is not defined, so f cannot be continuous at $x = 0$.

   COMMENT:

   You could ask students why the limit does not exist in choice (c).

6. Definition of continuity of a function at a point: *If the limit of the function exists at a point and is equal to the function evaluated at that point, then the function is continuous at that point.* Which of the following is not continuous at $x = c$?

   (a) (I) only
   (b) (II) only
   (c) (III) only
   (d) (I) and (II)
   (e) (I) and (III)
   (f) (II) and (III)

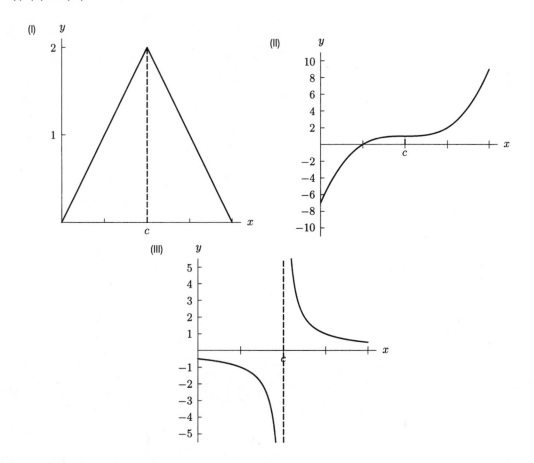

ANSWER:

(c)

COMMENT:

You could have students give reasons why the other choices are continuous functions.

Chapter Two

ConcepTests and Answers and Comments for Section 2.1

For Problems 1–3, assume your car has a broken speedometer.

1. In order to find my average velocity of a trip from Tucson to Phoenix, I need

    (a) The total distance of the trip
    (b) The highway mile markers
    (c) The time spent traveling
    (d) How many stops I made during the trip
    (e) A friend with a stop watch
    (f) A working odometer
    (g) None of the above

    ANSWER:

    (a) and (c)

    COMMENT:

    The choices are intentionally vague. This is meant to provide discussion. Your students may select more than one item.

2. In order to find my velocity at the instant I hit a speed trap, I need

    (a) The total distance of the trip
    (b) The highway mile markers
    (c) The time spent traveling
    (d) How many stops I made during the trip
    (e) A friend with a stop watch
    (f) A working odometer
    (g) none of the above

    ANSWER:

    (e) and (f). After I pass the speed trap I can watch my odometer as it increases by 0.1 miles while my friend (simultaneously) records the time it took to travel 0.1 miles.

    COMMENT:

    Using (e) and (f) as the odometer increases by 0.1 is a good estimate of the velocity at an instant. It may be beneficial to point out that if the odometer measured in hundredths of a mile, then you could compute an even better estimate of the instantaneous velocity.

3. Regarding the speed trap in Problem 2, when should your friend first start the stopwatch?

    (a) When the driver of an oncoming vehicle warns you of the speed trap ahead by flashing his/her bright headlights
    (b) When you spot the cop
    (c) Either scenario
    (d) Neither scenario

    ANSWER:

    (c). You can use an estimation of the average velocity before (or after) you hit the speed trap to estimate your actual velocity.

    COMMENT:

    The focus of this discussion should be on how h can be either positive or negative in order to estimate the derivative.

4. Which graph represents an object that is slowing down where t is time and D is distance. Assume the units on each axis is the same for all graphs.

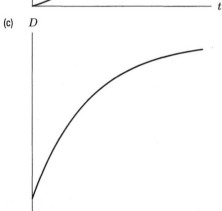

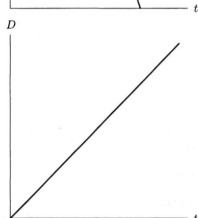

ANSWER:
(c)
COMMENT:
Students have a good idea of what the words "slowing down" mean, but try to make the connection between the words and slope of the tangent line at different points.

5. At approximately which integer value of x does the graph in Figure 2.1 have each of the following slopes?

   (a) -2

   (b) -1

   (c) 1

   (d) 2

   (e) 7

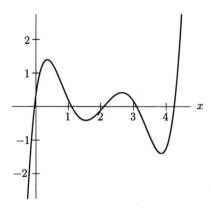

Figure 2.1

ANSWER:

(a) $x = 1$

(b) $x = 3$

(c) $x = 2$

(d) $x = 4$

(e) $x = 0$

COMMENT:

An enlarged version of this figure will make it easier to estimate slopes.

Follow-up Question. Put the slopes of the tangent lines occurring at $x = 0.5, 1.5, 2.5$, and 3.5 in order from smallest to largest.

Answer. $x = 3.5, x = 0.5, x = 1.5, x = 2.5$

6. For the graph of $y = f(x)$ in Figure 2.2 arrange the following numbers in ascending order (i.e. smallest to largest).

    (a) Slope of the graph where $x = 0.2$

    (b) Slope of the graph where $x = 1.5$

    (c) Slope of the graph where $x = 1.9$

    (d) Slope of the line connecting the points on the graph where $x = 1.5$ and $x = 1.9$

    (e) The number 1

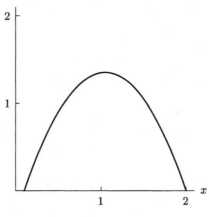

Figure 2.2

ANSWER:

(c), (d), (b), (e), (a)

COMMENT:

This is a good question for an elimination question in a classroom quiz session. One purpose for this question is to note the relationships between the slopes at the points $x = 1.5$ and $x = 1.9$ and the slope of the corresponding secant line.

7. Which of the following graphs represents the position of an object that is speeding up and then slowing down?

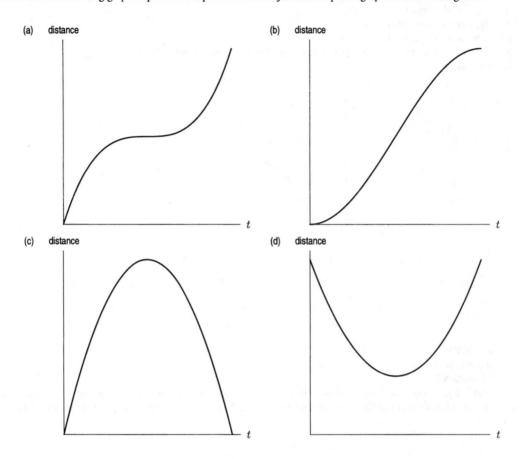

ANSWER:

(b). The instantaneous velocity is the slope of the curve at a point. Speeding up and slowing down requires the slope to increase and then decrease.

COMMENT:

This question is the same as Problem 25 in Section 1 of Chapter 1, but now you can relate the instantaneous velocity to the slope of the tangent line.

8. Which of the following graphs represents the position of an object that is slowing down?

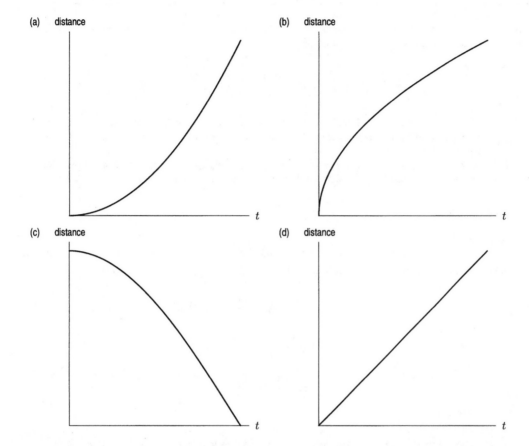

(a) distance

(b) distance

(c) distance

(d) distance

ANSWER:

(b). The instantaneous velocity is the slope of the curve at a point. For an object slowing down this means that the slope decreases and becomes closer to zero as time increases.

COMMENT:

You could have the students describe possible motion scenarios for the other choices. This question is the same as Problem 25 in Section 1 of Chapter 1.

ConcepTests and Answers and Comments for Section 2.2 ────

For Problems 1–2, we want to find how the volume, V, of a balloon changes as it is filled with air. We know $V(r) = \frac{4}{3}\pi r^3$, where r is the radius in inches and $V(r)$ is in cubic inches.

1. The expression

$$\frac{V(3) - V(1)}{3 - 1}$$

represents

   (a) The average rate of change of the radius with respect to the volume when the radius changes from 1 inch to 3 inches.
   (b) The average rate of change of the radius with respect to the volume when the volume changes from 1 cubic inch to 3 cubic inches.
   (c) The average rate of change of the volume with respect to the radius when the radius changes from 1 inch to 3 inches.
   (d) The average rate of change of the volume with respect to the radius when the volume changes from 1 cubic inch to 3 cubic inches.

   ANSWER:
   (c)
   COMMENT:
   This is a nice way for students to see the formula and verbal description for average rate of change together.

2. Which of the following represents the rate at which the volume is changing when the radius is 1 inch?

   (a)
   $$\frac{V(1.01) - V(1)}{0.01} = 12.69 \text{ in}^3$$

   (b)
   $$\frac{V(0.99) - V(1)}{-0.01} = 12.44 \text{ in}^3$$

   (c)
   $$\lim_{h \to 0} \left(\frac{V(1+h) - V(1)}{h} \right) \text{ in}^3$$

   (d) All of the above

   ANSWER:
   (d). Note that (c) is the exact rate of change while (a) and (b) approximate the rate of change.
   COMMENT:
   Students should be aware that (b) is equivalent to $\dfrac{V(1) - V(0.99)}{0.01}$.

3. For the function $g(x)$ shown in Figure 2.3, arrange the following numbers in increasing order.

   (a) 0   (b) $g'(-2)$ (c) $g'(0)$ (d) $g'(1)$ (e) $g'(3)$

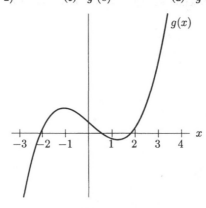

Figure 2.3

   ANSWER:
   (c), (d), (a), (b), (e)
   COMMENT:
   This can be used as an elimination question in a classroom quiz session.

4. Which of the following expressions represents the slope of a line drawn between the two points marked in Figure 2.4 ?

(a) $m = \dfrac{F(a) + F(b)}{a + b}$

(b) $m = \dfrac{F(b) - F(a)}{b - a}$

(c) $m = \dfrac{a}{b}$

(d) $m = \dfrac{F(a) - F(b)}{b - a}$

(e) $m = \dfrac{F(a) - F(b)}{a - b}$

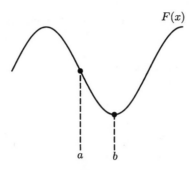

Figure 2.4

ANSWER:

(b) and (e). The coordinates of the two points shown are $(a, F(a))$ and $(b, F(b))$, so the slope of the line connecting them is $\dfrac{F(b) - F(a)}{b - a} = \dfrac{F(a) - F(b)}{a - b}$.

COMMENT:

You could repeat this question drawing the graph of a function that was increasing between a and b.

5. Which of the following expressions represents the slope of a line drawn between the two points marked in Figure 2.5?

(a) $\dfrac{F(\Delta x) - F(x)}{\Delta x}$

(b) $\dfrac{F(x + \Delta x) - F(x)}{\Delta x}$

(c) $\dfrac{F(x + \Delta x) - F(x)}{x}$

(d) $\dfrac{F(x + \Delta x) - F(x)}{x + x - \Delta x}$

(e) $\dfrac{F(x + \Delta x) - F(x)}{x + \Delta x}$

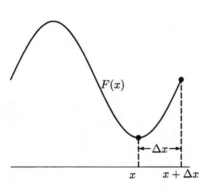

Figure 2.5

ANSWER:

(b). The coordinates of the two points are $(x, F(x))$ and $(x + \Delta x, F(x + \Delta x))$, so the slope of the line connecting them is

$$m = \frac{F(x + \Delta x) - F(x)}{x + \Delta x - x} = \frac{F(x + \Delta x) - F(x)}{\Delta x}.$$

COMMENT:

You could repeat this question drawing the graph of a function that was decreasing between x and $x + \Delta x$.

6. Assume that f is an odd function and that $f'(2) = 3$, then $f'(-2) =$

  (a) 3
  (b) -3
  (c) $1/3$
  (d) $-1/3$

   ANSWER:

   (a). This is best observed graphically.

   COMMENT:

   For a strong calculus class, have the students prove the result algebraically.

   Follow-up Question. If f is an even function and $f'(2) = 3$, then what is the value of $f'(-2) =$.
   Answer. $f'(2) = -3$.

7. Let $f(x) = x|x|$. Then $f(x)$ is differentiable at $x = 0$.

  (a) True
  (b) False

   ANSWER:

   (a). $f'(0) = \lim\limits_{h \to 0} \left(\dfrac{f(h) - f(0)}{h} \right)$. So $\lim\limits_{h \to 0^+} \dfrac{h|h|}{h} = \lim\limits_{h \to 0^+} \dfrac{h^2}{h} = 0$ and $\lim\limits_{h \to 0^-} \dfrac{h|h|}{h} = \lim\limits_{h \to 0^-} \dfrac{-h^2}{h} = 0$. Since the two limits exist and are equal, then $f(x)$ is differentiable at $x = 0$.

   COMMENT:

   Students often associate a minus sign with a number less than 0 rather than a number multiplied by -1.

ConcepTests and Answers and Comments for Section 2.3

1. Which of the following graphs (a)–(d) could represent the slope at every point of the function graphed in Figure 2.6?

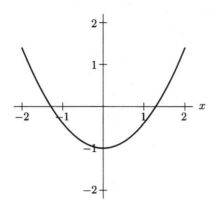

Figure 2.6

(a)

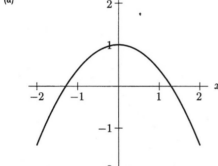

(b)

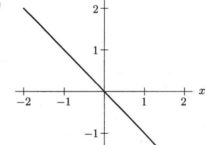

(c)

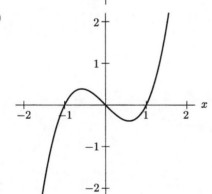

(d)

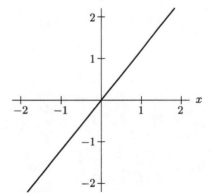

ANSWER:
(d)
COMMENT:
Your students should be analyzing where the extreme values of the original function occur and then whether the function is increasing or not. Have the students give specific reasons as to why the other three graphs could not represent the slope at every point of the function graphed in Figure 2.6.

2. Which of the following graphs (a)–(d) could represent the slope at every point of the function graphed in Figure 2.7?

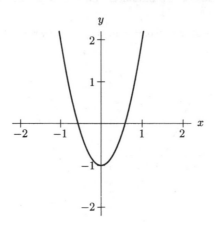

Figure 2.7

(a)
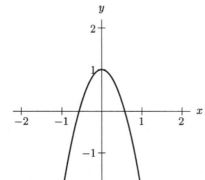

(b)
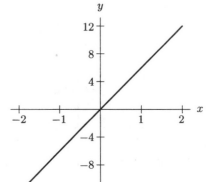

(c)
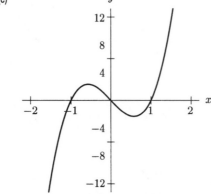

(d)
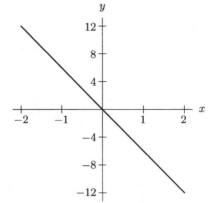

ANSWER:
(b). The function has negative slopes for $x < 0$, positive slopes for $x > 0$, and a zero slope for $x = 0$.
COMMENT:
You could have students explain why (a), (c), and (d) fail to be the correct answer.

3. Which of the following graphs (a)–(d) could represent the slope at every point of the function graphed in Figure 2.8?

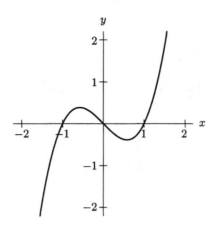

Figure 2.8

(a)

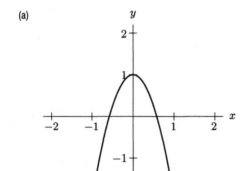

(b)

(c)

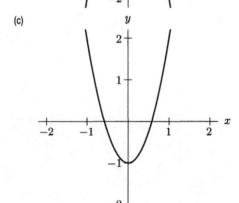

(d)

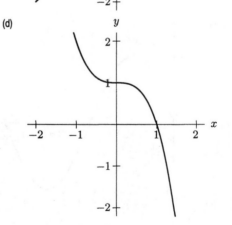

ANSWER:

(c). The function in Figure 2.8 has horizontal tangents at about $x = \pm 0.7$, is increasing for $x < -0.7$ and $x > 0.7$. Thus the graph of the slope of the function will be positive for $|x| > 0.7$, negative for $|x| < 0.7$, and zero for $x = \pm 0.7$.

COMMENT:

You could have students explain why (a), (b), and (d) fail to be the correct answer.

4. Which of the following graphs (a)–(d) could represent the slope at every point of the function graphed in Figure 2.9?

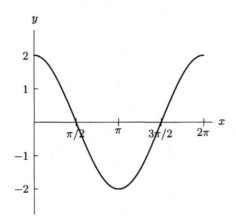

Figure 2.9

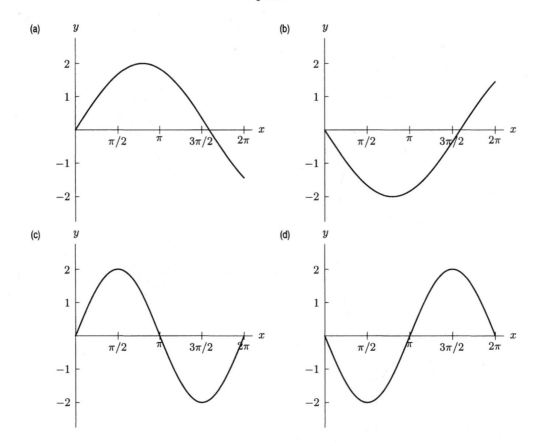

ANSWER:

(d). The original function has horizontal tangents at $0, \pi, 2\pi$, so the derivative will be zero there. This eliminates choices (a) and (b). The original function is also decreasing for $0 < x < \pi$, so its derivative must be negative in this region.

COMMENT:

You could have students explain why (a), (b), and (c) fail to be the correct answer.

5. Which of the following graphs (a)–(d) could represent the slope at every point of the function graphed in Figure 2.10?

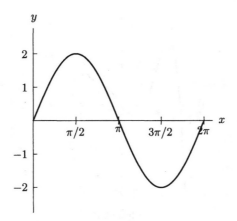

Figure 2.10

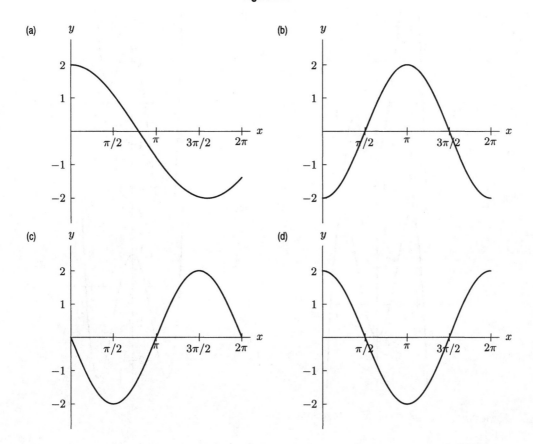

ANSWER:

(d). The original function has horizontal tangents at $x = \pi/2$ and $3\pi/2$, so the slope of the function is zero at those points. This eliminates (a) and (c). It is also increasing for $0 < x < \pi/2$, so its slope of the function is positive in this region.

COMMENT:

You could have students explain why (a), (b), and (c) fail to be the correct answer.

6. Which of the following graphs (a)–(d) could represent the slope at every point of the function graphed in Figure 2.11?

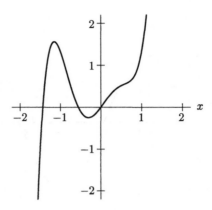

Figure 2.11

(a)
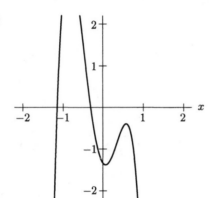

(b)
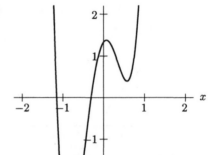

(c)
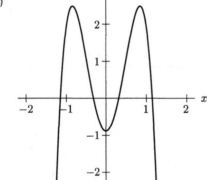

(d)
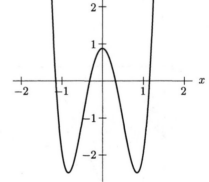

ANSWER:
(b)
COMMENT:
   Your students should analyze where the extreme values of the original function occur and where the function is increasing. Have the students give specific reasons as to why the other three graphs could not represent the slope at every point of the function graphed in Figure 2.11.

7. Suppose

$$f'(x) < 0, \text{ for } 0 < x < 2, \text{ for } 4 < x < 5, \text{ and for } 6 < x.$$

$$f'(x) > 0, \text{ for } x < 0, \text{ for } 2 < x < 4, \text{ and for } 5 < x < 6.$$

Which of the graphs (a)–(d) could be the graph of $f(x)$?

(a)
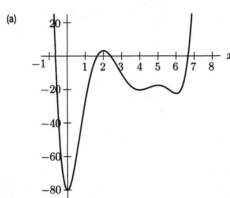

(b)
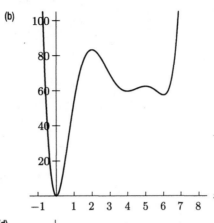

(c)
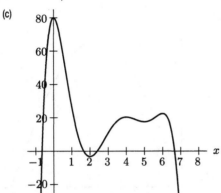

(d)
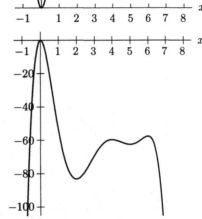

ANSWER:

(c) and (d)

COMMENT:

Your students should try to draw other graphs that meet the same requirements. Have students give reasons as to why the other two choices are invalid.

8. Which of the following graphs (a)–(d) could represent the function whose slope at every point is graphed in Figure 2.12?

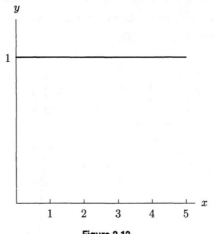

Figure 2.12

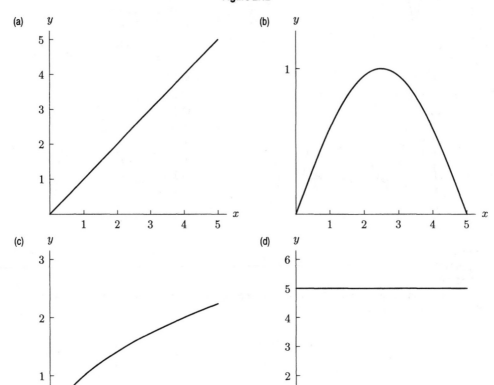

ANSWER:
(a). Because Figure 2.12 shows a constant value of 1, the original function will be a line with slope 1.
COMMENT:
This question (and the ones following) are meant to help students realize that they can think about the slope of a function from two viewpoints, either first using the function to find the slope at each point, or being given the slope at every point and then trying to recover the function. This should help them when they begin studying antiderivatives.

9. Which of the following graphs (a)–(d) could represent the function whose slope at every point is graphed in Figure 2.13?

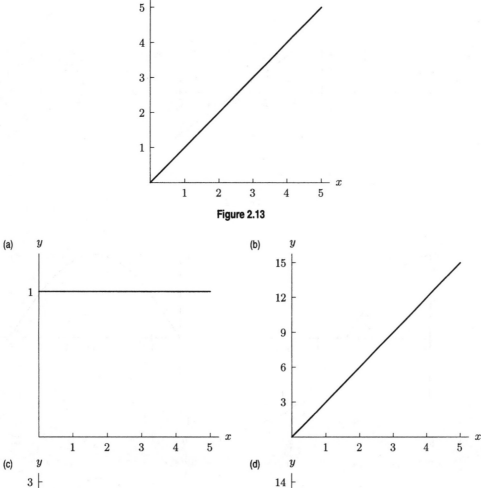

Figure 2.13

(a)

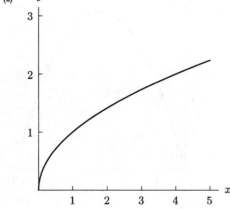

(b)

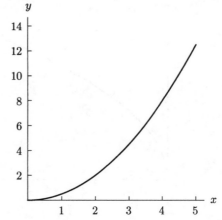

(c)

(d)

ANSWER:

(d). The graph of the slope is positive, and increasing from 0 to 5 as x goes from 0 to 5, therefore the function must be increasing and concave up.

COMMENT:

See Comment for Problem 8.

10. Which of the following graphs (a)–(d) could represent the function whose slope at every point is graphed in Figure 2.14?

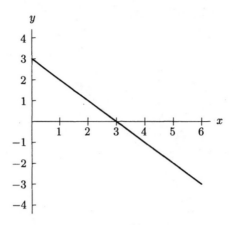

Figure 2.14

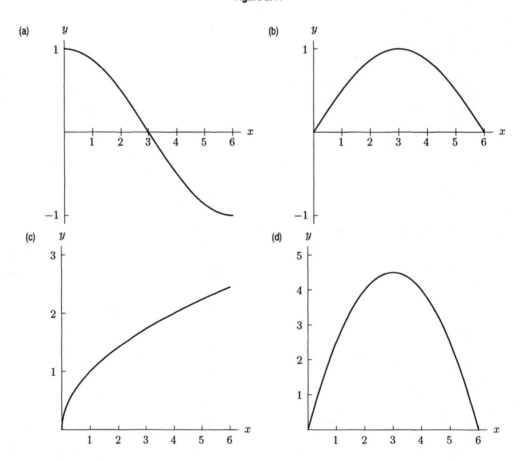

ANSWER:

(d). Since the graph of the slope is decreasing at every point, the function will be concave down. The graph of the function needs an initial slope of 3 and a horizontal tangent at $x = 3$.

COMMENT:

See Comment for Problem 8.

11. Which of the following graphs (a)–(d) could represent the function whose slope at every point is graphed in Figure 2.15?

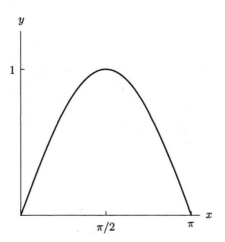

Figure 2.15

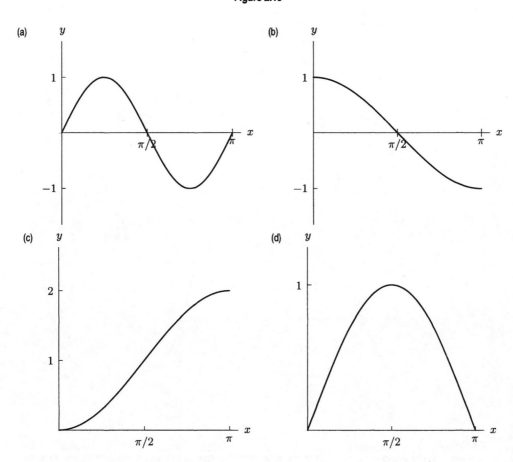

ANSWER:

(c). The graph of the slope is positive at every point, so the function must be increasing.

COMMENT:

See Comment for Problem 8.

12. Which of the following graphs (a)–(d) could represent the function whose slope at every point is graphed in Figure 2.16?

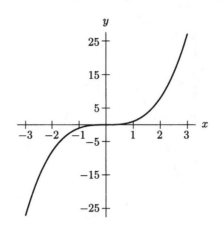

Figure 2.16

(a)

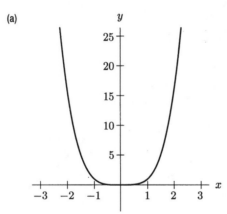

(b)

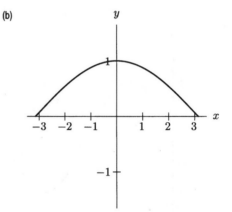

(c)

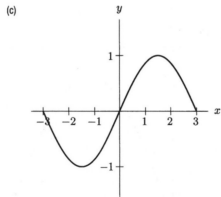

(d)

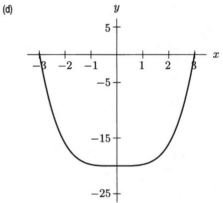

ANSWER:

(a) and (d). The graph of the slope at every point increases from a negative value at $x = -3$, to a slope of zero at $x = 0$, to a positive value for $x > 0$.

COMMENT:

See Comment for Problem 8.

13. Which of the following is a graph of a function that is equal to its own derivative, that is, $f'(x) = f(x)$.

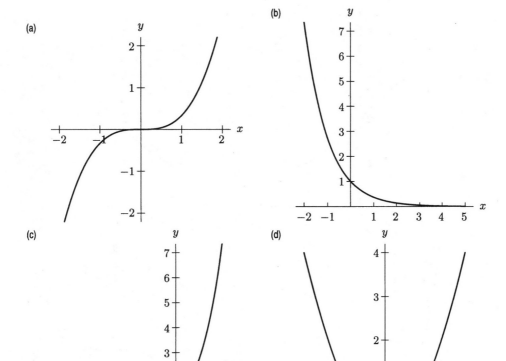

(a)

(b)

(c)

(d)

ANSWER:

(c). The slope of the curve in (a) is always positive while the function has both positive and negative values, so $f'(x) \neq f(x)$. The slope of the curve in (b) is always negative, while the function has only positive values, so $f'(x) \neq f(x)$. In (c), both the function and its slope are positive, and the slope at $(0, 1)$ appears to be 1. The function and slope increase together, so $f'(x) = f(x)$. The graph in (d) has only positive values while its slope has both positive and negative values, so $f'(x) \neq f(x)$.

COMMENT:

You could have students suggest formulas for the graphs in each choice.

14. Which of the following is a graph of a function that is equal to the negative of its own derivative, that is, $f(x) = -f'(x)$.

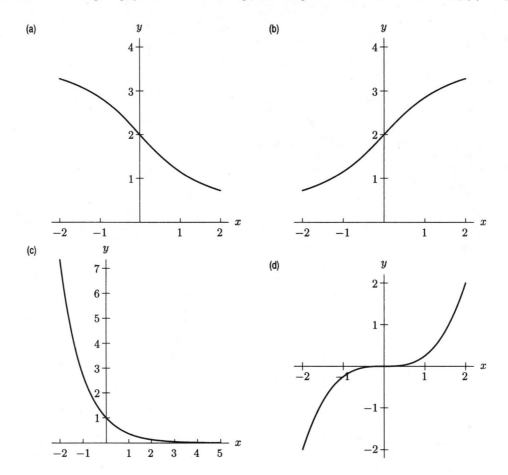

ANSWER:

(c). With $f(x) = -f'(x)$, the function will have positive slopes where it is negative and negative slopes where it is positive. This rules out (b) and (d). With $f(x) = -f'(x)$, as the function gets larger, the slope will be steeper. This rules out (a). The slope in (c) at $(0, 1)$ appears to be -1.

COMMENT:

You could have students suggest formulas for the graphs in each choice.

15. Which of the following graphs (a)–(d) could represent the slope at every point of the function graphed in Figure 2.17?

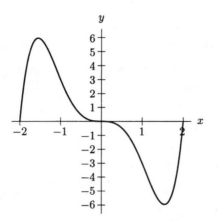

Figure 2.17

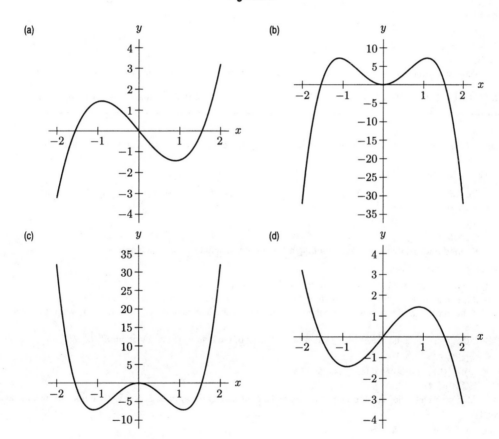

ANSWER:

(c). The graph of the function in Figure 2.17 has horizontal tangents at $x \approx 0, -1.6, 1.6$, is increasing for $x < -1.6$ and $x > 1.6$, and is decreasing for $-1.6 < x < 1.6$. Thus, the derivative will be positive for $x < -1.6$ and $x > 1.6$, negative for $0 < |x| < 1.6$, and zero for $x \approx 0, -1.6, 1.6$.

COMMENT:

You could have students explain why (a), (b), and (d) fail to be the correct answer.

16. If $f'(x) = g'(x)$, then $f(x) = g(x)$.

    (a) True

    (b) False

        ANSWER:

        (b). If $f'(x) = g'(x)$, then $f(x) = g(x) + C$, where C is some constant.

        COMMENT:

        You might point out that the graphs of f and g differ by a vertical shift. A student's instinct is to answer **True**. Have students suggest their own counterexamples.

17. If $y = \pi^5$, then $y' = 5\pi^4$.

    (a) True

    (b) False

        ANSWER:

        (b). Since π^5 is a constant, then $y' = 0$.

        COMMENT:

        This question seems remarkably obvious in class. However, on an exam students tend to miss this question.

18. If $y = (x + 1)(x + 2)(x + 3)(x + 4)$, then $\dfrac{d^5 y}{dx^5} = 0$.

    (a) True

    (b) False

        ANSWER:

        (a). y is a polynomial of degree four.

        COMMENT:

        Make sure your students realize that they don't need to take a derivative to answer this question.

ConcepTests and Answers and Comments for Section 2.4

For Problems 1–3, the function $C(r)$ is the total cost of paying off a car loan borrowed at an interest rate of $r\%$ per year.

1. What are the units of $C'(r) = \dfrac{dC}{dr}$?

    (a) Year/\$ (b) \$/Year   (c) \$/(\%/Year)   (d) (\%/Year)/\$

        ANSWER:

        (c)

        COMMENT:

        Remind students that many times looking at the units will help them solve the problem.

2. What is the practical meaning of $C'(5)$?

    (a) The rate of change of the total cost of the car loan is $C'(5)$.

    (b) If the interest rate increases by 1%, then the total cost of the loan increases by about $C'(5)$.

    (c) If the interest rate increases by 1%, then the total cost of the loan increases by about $C'(5)$ when the interest rate is 5%.

    (d) If the interest rate increases by 5%, then the total cost of the loan increases by about $C'(5)$.

        ANSWER:

        (c). $C'(5)$ requires the interest rate to be 5%.

        COMMENT:

        This problem is designed to help students make connections between real-world problems and mathematical concepts.

3. What is the sign of $C'(5)$?

    (a) Positive

    (b) Negative

    (c) Not enough information

        ANSWER:

        (a). If the interest rate increases, then the total car loan cost also increases, thus $C'(5)$ is positive.

        COMMENT:

        Students don't always realize how to make connections between real-world problems and mathematical concepts.

For Problems 4–5, you invest $1000 at an annual interest rate of $r\%$, compounded continuously. At the end of 10 years, you have a balance of B dollars, where $B = g(r)$.

4. What is the financial interpretation of $g(5) = 1649$?

   (a) When r is 5%, B is $1649.
   (b) When the interest rate is 5, you have 1649.
   (c) When the interest rate is 5%, there is $1649.
   (d) When the interest rate is 5%, then in 10 years you have a balance of $1649.

   ANSWER:
   (d) contains the most information.

   COMMENT:
   Try to get your students to see the subtle differences between the four choices.

5. What is the financial interpretation of $g'(5) = 165$?

   (a) The balance in your account after 5 years is $165
   (b) The balance grows at a rate of $165 per % when $r = 5\%$.
   (c) If the interest rate increases from 5% to 6%, you would expect about $165 more in your account.
   (d) If the interest rate increases from 5% to 6% you would expect about $1814 in your account.

   ANSWER:
   (c) and (d) are equivalent. (d) uses the information from the previous problem.

   COMMENT:
   Students should start thinking of rate as an incremental change.

For Problems 6–7, assume $g(v)$ is the fuel efficiency, in miles per gallon, of a car going at a speed of v miles per hour.

6. What are the units of $g'(v) = \dfrac{dg}{dv}$?

   (a) $(\text{miles})^2/(\text{gal})(\text{hour})$
   (b) hour/gal
   (c) gal/hour
   (d) $(\text{gal})(\text{hour})/(\text{miles})^2$

   ANSWER:
   (b)

   COMMENT:
   Students may find it more instructive not to cancel like terms.

7. What is the practical meaning of $g'(55) = -0.54$?

   (a) When the car is going 55 mph, the rate of change of the fuel efficiency decreases *to* approximately 0.54 miles/gal.
   (b) When the car is going 55 mph, the rate of change of the fuel efficiency decreases *by* approximately 0.54 miles/gal.
   (c) If the car speeds up from 55 mph to 56 mph, then the fuel efficiency is approximately -0.54 miles per gallon.
   (d) If the car speeds up from 55 mph to 56 mph, then the car becomes less fuel efficient by approximately 0.54 miles per gallon.

   ANSWER:
   (b) and (d) are equivalent, with (d) containing the most information. Notice that (a) and (c) are wrong.

   COMMENT:
   Students should strive to provide descriptions that are as complete as (d).

8. Let $N = f(t)$ be the total number of cans of cola Sean has consumed by age t in years. Interpret the following in practical terms, paying close attention to units.

   (a) $f(14) = 400$ (b) $f^{-1}(50) = 6$
   (c) $f'(12) = 50$ (d) $(f^{-1})'(450) = 1/70$

   ANSWER:

   (a) By age 14, Sean had consumed 400 cans of cola.
   (b) Sean had consumed 50 cans of cola by age 6.
   (c) At age 12 Sean was consuming approximately 50 cans of cola per year.
   (d) At the time Sean had consumed 450 cans of cola, it took approximately one year to consume 70 additional cans.

   COMMENT:
   Students have a difficult time with derivatives of inverses.

For Problems 9–11, let $A = f(t)$ be the depth of tread, in centimeters, on a radial tire as a function of the time elapsed t, in months, since the purchase of the tire.

9. Interpret the following in practical terms, paying close attention to units.

    (a) $f(6) = 0.5$ (b) $f^{-1}(0.31) = 15$
    (c) $f'(12) = -0.015$ (d) $(f^{-1})'(0.4) = -60$

    ANSWER:

    (a) After 6 months, the depth of tread was 0.5 cm.
    (b) When the depth of tread was 0.31 cm, the tire had been used for 15 months.
    (c) After 12 months, the depth of tread was decreasing by 0.015 cm per month.
    (d) When the depth of tread was 0.4 cm, it took 60 months to reduce the tread by 1 cm (or it took one month to reduce the depth of thread by $1/60$ cm).

    COMMENT:

    For a challenge have students construct a similar word problem.

10. What is the sign of $f'(t)$? Explain why.

    ANSWER:

    The sign of $f'(t)$ is negative, because tread wears with use, so its depth is always decreasing.

    COMMENT:

    Follow-up Question. What would $f'(t) > 0$ mean?

11. What is the sign of $(f^{-1})'(A)$? Explain why.

    ANSWER:

    $A = f(t)$ is a decreasing function, so its inverse will also be decreasing, and $(f^{-1})'(A)$ will be negative.

    COMMENT:

    This reasoning might be pointed out graphically.

ConcepTests and Answers and Comments for Section 2.5

1. The graph of $f(x)$ is shown in Figure 2.18. Which of the following are true for f as shown in this window?

    (a) $f(x)$ is positive   (b) $f(x)$ is increasing   (c) $f'(x)$ is positive
    (d) $f'(x)$ is increasing   (e) $f''(x)$ is non-negative

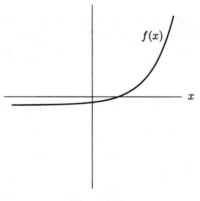

Figure 2.18

ANSWER:
(b), (c). (d), and (e)
COMMENT:
You could repeat this problem with other graphs.

2. If $f'(x)$ is positive, then $f''(x)$ is increasing.

   (a) True

   (b) False

      ANSWER:

      (b). $f'(x)$ positive means $f(x)$ is increasing. $f'(x) = x^4 - 8x^2 + 18$ provides a counterexample.

      COMMENT:

      Have students provide their own counterexample. You might also phrase this question in terms of concavity and give graphical counterexamples.

3. If $f'(x)$ is increasing, then $f(x)$ is increasing.

   (a) True

   (b) False

      ANSWER:

      (b). If $f'(x)$ is increasing, then the only acceptable conclusion is that $f(x)$ is concave up. For an example, consider $f'(x) = 2x$, then a possibility for $f(x)$ is x^2 which is not always increasing.

      COMMENT:

      Have students provide their own counterexample. You might also phrase this question in terms of concavity and give graphical counterexamples.

4. If $f''(x)$ is positive, then $f(x)$ is concave up.

   (a) True

   (b) False

      ANSWER:

      (a)

      COMMENT:

      You could ask what is true if $f''(x) < 0$.

5. If $f''(x)$ is positive, then $f'(x)$ is increasing.

   (a) True

   (b) False

      ANSWER:

      (a)

      COMMENT:

      You might note that $f''(x)$ is the rate of change of $f'(x)$.

6. If $f'(x)$ is increasing, then $f(x)$ is concave up.

   (a) True

   (b) False

      ANSWER:

      (a)

      COMMENT:

      You might note that $f'(x)$ increasing means $f''(x)$ is positive.

7. If the velocity of an object is constant, then its acceleration is zero.

   (a) True

   (b) False

      ANSWER:

      (a)

      COMMENT:

      Follow-up Question. If the velocity is zero at a specific instant in time, does the acceleration need to be zero at that same time also?

      Answer. No, a grapefruit that is tossed straight up in the air has a velocity of 0 ft/sec when the grapefruit reaches the highest point it will travel. However, at the point the acceleration of the grapefruit is that of gravity, which is not 0 ft/sec^2.

8. The value of the second derivative of the function shown in Figure 2.19 at the point $x = 1$ is

    (a)   Positive   (b)   Negative

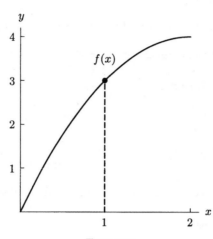

Figure 2.19

ANSWER:

(b). As x increases, the slope of the tangent line decreases. Thus the second derivative is not positive.

COMMENT:

You could ask students if the magnitude of the second derivative of a function can be determined from the graph of the function. It cannot. For example, consider the function $f(x) = x^2$. It looks almost straight in places, i.e. no concavity, which would imply that the second derivative is zero. But, the value of the second derivative is always 2.

9. In Figure 2.20, the second derivative at points a, b, and c is (respectively),

    (a) $+, 0, -$ (b) $-, 0, +$ (c) $-, 0, -$

    (d) $+, 0, +$ (e) $+, +, -$ (f) $-, -, +$

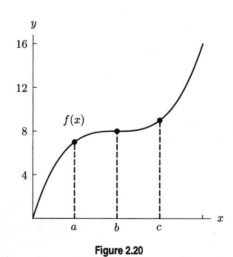

Figure 2.20

ANSWER:

(b). The graph is concave down at a, so $f''(a) \leq 0$ leaving (b), (c), and (f). The graph is concave up at c, so $f''(c) \geq 0$ leaving (b) and (f). The graph has an inflection point at b, so $f''(b) = 0$ leaving (b).

COMMENT:

See Problem 8.

10. In Figure 2.21, the second derivative at points a, b, and c is (respectively),

(a) $+, 0, -$ (b) $-, 0, +$ (c) $-, 0, -$
(d) $+, 0, +$ (e) $0, +, 0$ (f) $0, -, 0$

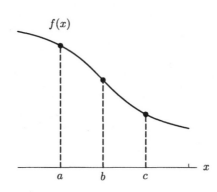

Figure 2.21

ANSWER:

(b). The graph is concave down at a, so $f''(a) \leq 0$ leaving (b), (c), (e), and (f). The graph is concave up at c, so $f''(c) \geq 0$ leaving (b), (e), and (f). The graph has an inflection point at b, so $f''(b) = 0$ leaving (b).

COMMENT:

See Problem 8.

11. In Figure 2.22, at $x = 0$ the signs of the function and the first and second derivatives, in order, are

(a) $+, 0, +$ (b) $+, 0, -$ (c) $-, +, -$
(d) $-, +, +$ (e) $+, -, +$ (f) $+, +, +$

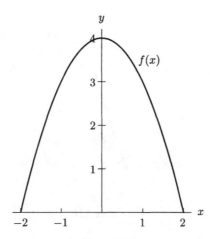

Figure 2.22

ANSWER:

(b). At $x = 0$ the graph is positive, has a horizontal tangent, and is concave down.

COMMENT:

See Problem 8.

12. In Figure 2.23, at $x = 0$ the signs of the function and the first and second derivatives, in order, are

(a) $-, +, +$ (b) $-, -, -$ (c) $-, +, -$
(d) $-, +, +$ (e) $+, -, +$ (f) $+, +, +$

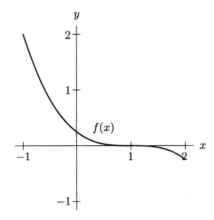

Figure 2.23

ANSWER:
(e). At $x = 0$ the graph is positive, decreasing, and concave up.
COMMENT:
See Problem 8.

13. In Figure 2.24, at $x = 0$ the signs of the function and the first and second derivatives, in order, are

(a) $+, 0, +$ (b) $-, 0, -$ (c) $+, 0, -$
(d) $-, +, 0$ (e) $+, -, 0$ (f) $+, +, +$

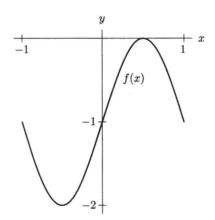

Figure 2.24

ANSWER:
(d). At $x = 0$ the graph is negative, increasing, and has an inflection point.
COMMENT:
See Problem 8.

14. Which of the following graphs (a)–(d) could represent the second derivative of the function in Figure 2.25 ?

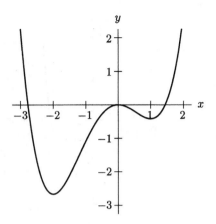

Figure 2.25

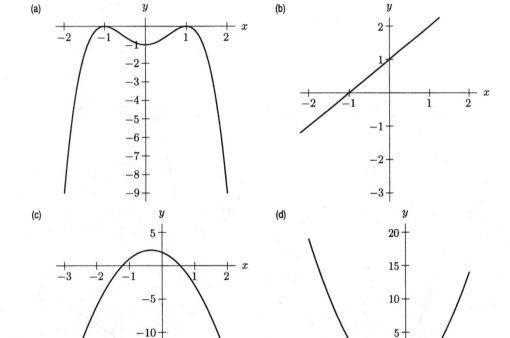

ANSWER:
(d). The graph in Figure 2.25 is concave up for $x < -1.2$ and $x > 0.5$ with inflection points at $x \approx -1.2$ and 0.5. It is concave down elsewhere. So the second derivative is positive for $x < -1.2$ and $x > 0.5$, negative for $-1.2 < x < 0.5$, and zero at $x \approx -1.2$ and 0.5.
COMMENT:
You could have students explain why (a), (b), and (c) fail to be the correct answer.

15. Which of the following graphs (a)–(d) could represent the second derivative of the function in Figure 2.26?

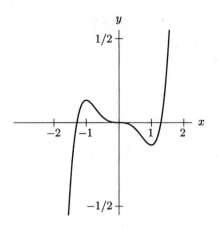

Figure 2.26

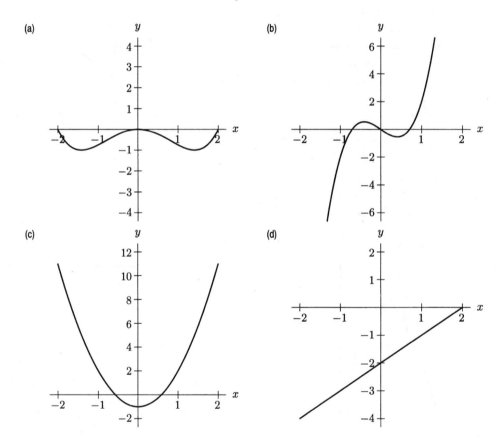

ANSWER:

(b). The graph in Figure 2.26 appears to be concave down for $-2 < x < -0.7$ and $0 < x < 0.7$. It is concave up elsewhere with inflection points at $x \approx -0.7, 0,$ and 0.7.

COMMENT:

You could have students explain why (a), (c), and (d) fail to be the correct answer.

16. Figure 2.27 shows position as a function of time for two sprinters running in parallel lanes. Which of the following is true?

    (a) At time A, both sprinters have the same velocity.
    (b) Both sprinters continually increase their velocity.
    (c) Both sprinters run at the same velocity at some time before A.
    (d) At some time before A, both sprinters have the same acceleration.

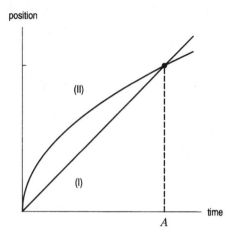

Figure 2.27

ANSWER:

(c). The sprinter whose position is given by (I) has a constant velocity, represented by the slope of the line. Since the slope of the curve (II) continually decreases, the velocity of the sprinter is continually decreasing. At A both sprinters have the same position. The acceleration for sprinter (I) is zero, so the only true statement is (c). They have the same velocity when the slope of curve (II) is parallel with the line (I).

COMMENT:

You might point out the relationship between this problem and the Mean Value Theorem.

17. If an object's acceleration is negative, at that particular instant the object can be

    (a) Slowing down only
    (b) Speeding up only
    (c) Slowing down or momentarily stopped only
    (d) Slowing down, momentarily stopped, or speeding up

    ANSWER:

(d). The acceleration of an object is the rate of change of its velocity with respect to time. If the acceleration is negative, its velocity is decreasing, but this tells us nothing about the value of the velocity.

COMMENT:

You could have students provide position graphs of an object with negative acceleration which satisfies (a), (b), and (c), respectively.

18. Figure 2.28 shows the graph of position versus time, t. Which of (a)–(d) represents a corresponding graph of acceleration as a function of time?

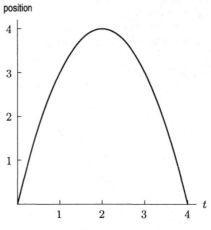

position

Figure 2.28

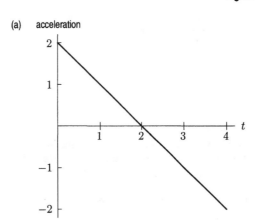

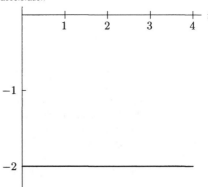

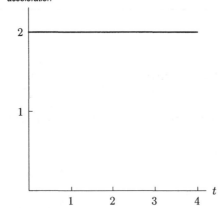

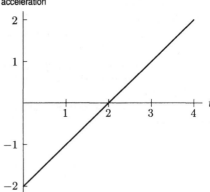

ANSWER:
(b). The position graph is concave down for $0 < t < 4$. Thus the acceleration is not positive for $0 < t < 4$.
COMMENT:
You could have students give specific points on the graphs of the other choices which have properties that are not consistent with Figure 2.28.

19. Figure 2.29 shows the graph of position versus time, t. Which of (a)–(d) represents a corresponding graph of acceleration as a function of time?

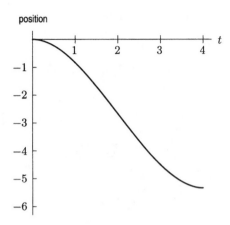

position

Figure 2.29

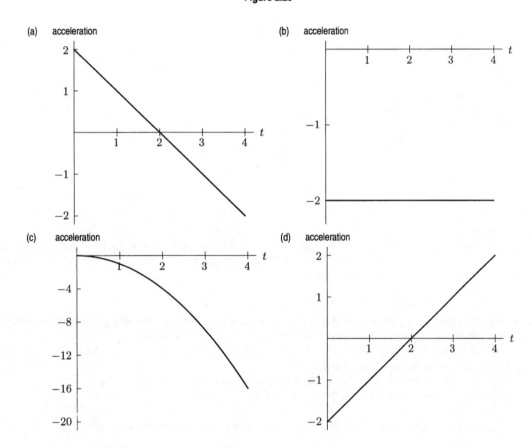

(a) acceleration

(b) acceleration

(c) acceleration

(d) acceleration

ANSWER:

(d). The graph appears to be concave down for $0 < t < 2$, concave up for $2 < t < 4$ with an inflection point at $t = 2$. Thus the acceleration is not positive for $0 < t < 2$, is not negative for $2 < t < 4$, and is zero at $t = 2$.

COMMENT:

You could have students give specific points on the graphs of the other choices which have properties that are not consistent with Figure 2.29.

20. Figure 2.30 represents acceleration as a function of time, t. Which of the following could represent the corresponding position versus time graph?

(a) (I)

(b) (II)

(c) (III)

(d) (I) and (II)

(e) (I), (II), and (III)

(f) None of these

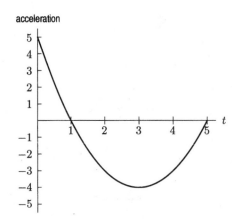

acceleration

Figure 2.30

(I)

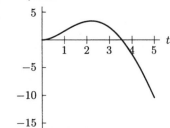

position

(II)

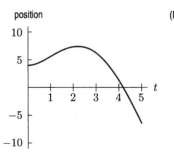

position

(III)

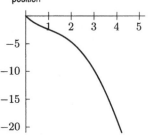

position

ANSWER:

(e). From Figure 2.30 we notice that the graph of the position function is concave up for $0 < t < 1$, is concave down for $1 < t < 5$, and has an inflection point when $t = 1$. Since the graphs shown in (I), (II), and (III) have these properties, then each could be a possible graph of the position function.

COMMENT:

You might point out that the graphs in (I) and (II) differ by a vertical translation.

21. At a specific instant in time we observed that the distance scale of the universe was increasing. For all time we can prove that the second derivative of the distance scale with respect to time is always negative.

Which of the following is true?

(a) The universe will keep expanding forever.

(b) At some point in the future the universe will stop expanding and begin contracting.

(c) With the given information either of these is a possibility.

ANSWER:

(c). A negative second derivative is possible for functions which are either increasing or decreasing.

COMMENT:

Follow-up Question. What if we knew that the distance scale was always increasing. Would that change the answer? **Answer.** Yes, with the additional information we know that the first derivative is always positive. Therefore (a) is now correct. However, it does not say the distance scale grows without bound. It may simply approach an asymptote.

22. In *Star Trek: First Contact*, Worf almost gets knocked into space by the Borg. Assume he was knocked into space and his space suit was equipped with thrusters. Worf fires his thruster for 1 second which produces a constant acceleration in the positive direction. In the next second he turns off his thrusters. In the third second he fires his thrusters producing a constant negative acceleration. The acceleration as a function of time is given in Figure 2.31. Which of (a)–(d) represent his position versus time graph?

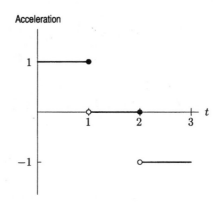

Figure 2.31

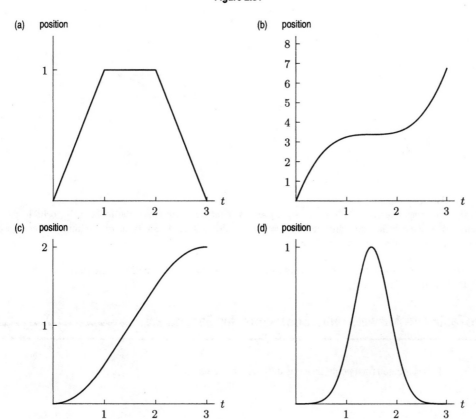

ANSWER:

(c). From the acceleration graph we see that the position graph will be concave up for $0 < t < 1$, concave down for $2 < t < 3$ and have a constant slope for $1 < t < 2$.

COMMENT:

You could have students give specific points on the graphs in the other choices which have properties not consistent with given acceleration graph.

23. Which of the following graphs satisfies the relationship $f''(x) = -f(x)$?

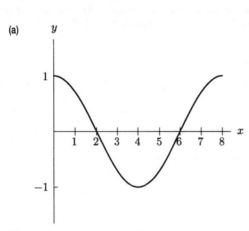

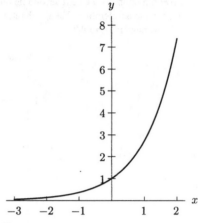

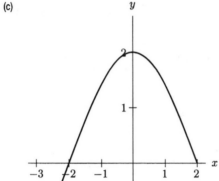

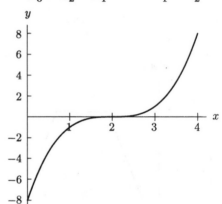

ANSWER:

(a). Functions that satisfy $f''(x) = -f(x)$ will be concave down where the function is positive and concave up where it is negative. Inflection points occur where the function is zero. The answer (c) would also be correct if we could tell that inflection points occurred at $x = \pm 2$.

COMMENT:

You could have students give specific points on the graphs in the other choices which have properties not consistent with the fact that $f''(x) = -f(x)$.

ConcepTests and Answers and Comments for Section 2.6

1. If a function is continuous at a point, then it is differentiable at that point.

(a) True
(b) False

ANSWER:

(b). For example, the function $f(x) = |x|$ is continuous at $x = 0$, but it is not differentiable there.

COMMENT:

You might have students give other examples. ($f(x) = x^{1/3}$ is one.)

Chapter Three

ConcepTests and Answers and Comments for Section 3.1 ━━━━━━━

1. The graph of a function f is given in Figure 3.1. If f is a polynomial of degree 3, then the value of $f'''(0)$ is

   (a) Positive
   (b) Negative
   (c) Zero

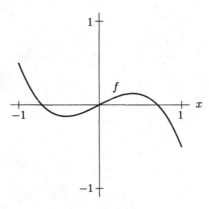

Figure 3.1

ANSWER:

(b). Because the graph of this polynomial of degree 3 is negative for large values of x, the coefficient of x^3 will be negative. (Recall the third derivative of a polynomial of degree 3 is a constant.)

COMMENT:

You could ask students why $f(x)$ could not become positive for $x > 1$.

2. The graph of a function f is given in Figure 3.2. If f is a polynomial of degree 3, then $f'''(0)$ is

   (a) Positive
   (b) Negative
   (c) Zero

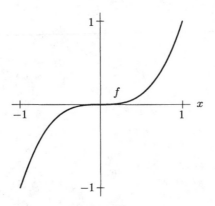

Figure 3.2

ANSWER:

(a). Because the graph of this polynomial of degree 3 is positive for large values of x, the coefficient of x^3 will be positive. (Recall the third derivative of a polynomial of degree 3 is a constant.)

COMMENT:

You could ask students if there could be other inflection points for f.

3. The graph of a function f is given in Figure 3.3. If f is a polynomial of degree 3, then the values of $f'(0)$, $f''(0)$, and $f'''(0)$ are (respectively)

(a) $0, 0, +$
(b) $0, 0, -$
(c) $0, +, -$
(d) $0, -, -$
(e) $+, -, +$
(f) $0, +, +$

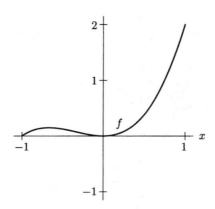

Figure 3.3

ANSWER:

(f). There is a horizontal tangent at the origin, so $f'(0) = 0$. The graph shows that f has horizontal intercepts at -1 and 0, with a double root at 0. Thus f has the form $f(x) = k(x+1)x^2$. Because $f(x) > 0$ for $x > 0$, then $k > 0$. So $f'(x) = k(3x^2 + 2x)$, $f''(x) = k(6x+2)$, and $f'''(x) = 6k$.

COMMENT:

You could ask students why a double root at zero means that f has a factor of x^2.

4. The graph of a function f is given in Figure 3.4. If f is a polynomial of degree 3, then the values of $f'(0)$, $f''(0)$, and $f'''(0)$ are (respectively)

(a) $+, 0, +$
(b) $-, 0, -$
(c) $+, 0, -$
(d) $+, -, -$
(e) $+, -, +$
(f) $+, +, +$

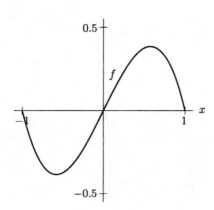

Figure 3.4

ANSWER:

(c). The graph shows that f has horizontal intercepts at $x = -1, 0$, and 1. Thus f has the form $f(x) = kx(x+1)(x-1)$. Because $f(x) > 0$ for $0 < x < 1$, we have $k < 0$. Then $f'(x) = k(3x^2 - 1)$, $f''(x) = k(6x)$, and $f'''(x) = 6k$.

COMMENT:

You could ask why there could not be another horizontal intercept outside this viewing window.

5. The graph of a function f is given in Figure 3.5. If f is a polynomial of degree 3, then the values of $f'(0)$, $f''(0)$, and $f'''(0)$ are (respectively)

(a) $-,+,+$ (b) $-,-,-$ (c) $-,+,-$

(d) $-,+,+$ (e) $+,-,+$ (f) $+,+,+$

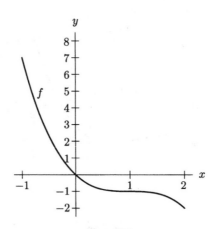

Figure 3.5

ANSWER:

(c). At $x = 0$, the graph is decreasing and concave up, so $f'(0) < 0$ and $f''(0) \geq 0$. Because the graph becomes more negative as x increases beyond 1, the sign of the coefficient of x^3 (and thus the sign of $f'''(0)$) is negative. (Recall the third derivative of a polynomial of degree 3 is a constant.)

COMMENT:

You could ask why the graph of this function could not become positive for larger values of x.

6. The graph of a function f is given in Figure 3.6. If f is a polynomial of degree 3, then the values of $f'(0)$, $f''(0)$, and $f'''(0)$ are (respectively)

(a) $-,-,+$ (b) $-,0,-$ (c) $-,+,-$

(d) $-,+,+$ (e) $+,-,+$ (f) $+,+,+$

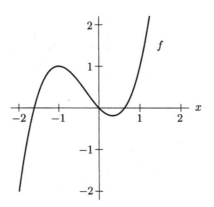

Figure 3.6

ANSWER:

(d). At $x = 0$, the graph is decreasing and concave up, so $f'(0) < 0$ and $f''(0) \geq 0$. Because the graph is positive as x increases beyond 0.5, the sign of the coefficient of x^3 (and thus the sign of $f'''(0)$) is positive. (Recall the third derivative of a polynomial of degree 3 is a constant.)

COMMENT:

You could ask why the graph of this function could not become negative for larger values of x.

ConcepTests and Answers and Comments for Section 3.4 ▬▬▬▬

1. Given the graphs of the functions $f(x)$ and $g(x)$ in Figures 3.7 and 3.8, which of (a)–(d) is a graph of $f(g(x))$?

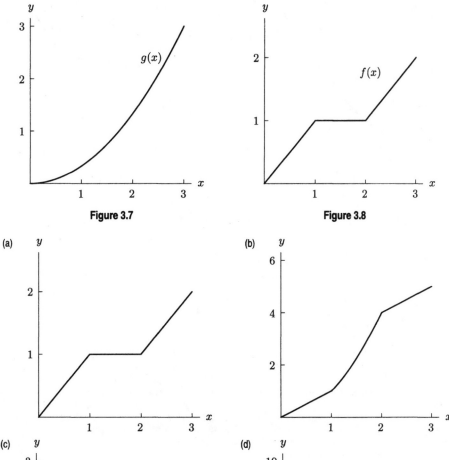

Figure 3.7   Figure 3.8

ANSWER:

(c). Because $(f(g(x)))' = f'(g(x))g'(x)$, we see $f(g(x))$ has a horizontal tangent whenever $g'(x) = 0$ or $f'(g(x)) = 0$. Now, $f'(g(x)) = 0$ for $1 < g(x) < 2$ and this approximately corresponds to $1.7 < x < 2.5$.

COMMENT:

You could have students give specific places in the other choices where there was a conflict with information given in the graphs of f and g.

2. Given the graphs of the functions $f(x)$ and $g(x)$ in Figures 3.9 and 3.10, which of (a)–(d) is a graph of $f(g(x))$?

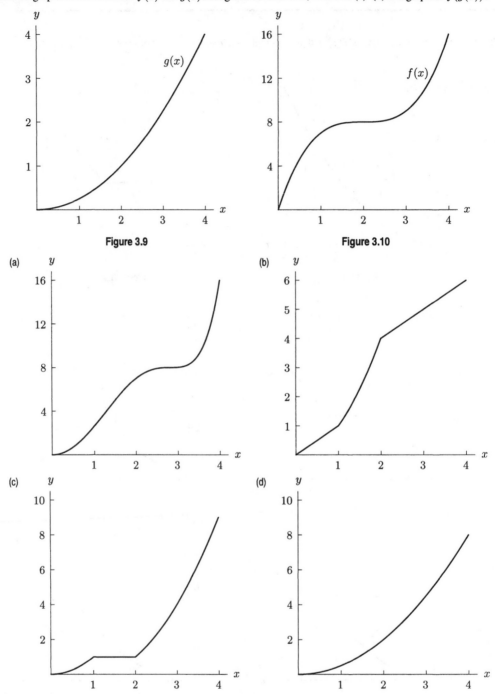

Figure 3.9

Figure 3.10

ANSWER:

(a). Because $(f(g(x)))' = f'(g(x))g'(x)$, we see $f(g(x))$ has a horizontal tangent whenever $g'(x) = 0$ or $f'(g(x)) = 0$. Now $f'(x) = 0$ only when $x = 2$, so the composite function only has horizontal tangents when $g'(x) = 0$ or when $g(x) = 2$.

COMMENT:

Students may want to check points, which is tedious. This is an opportunity to show the power of reasoning based on the chain rule.

3. Given the graphs of the function $g(x)$ and $f(x)$ in Figures 3.11 and 3.12 , which of (a)–(d) represents $f(g(x))$?

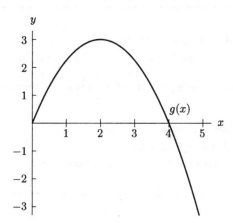

Figure 3.11

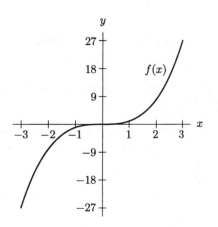

Figure 3.12

(a)

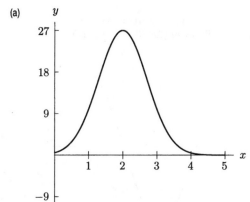

(b)

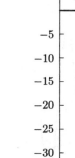

(c)

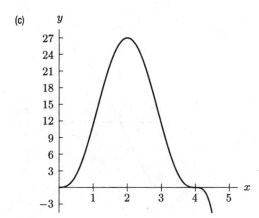

(d)

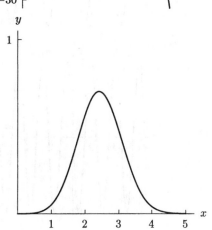

ANSWER:

(c). Because $(f(g(x)))' = f'(g(x))g'(x)$, we see $f(g(x))$ has a horizontal tangent whenever $g'(x) = 0$ or $g(x) = 0$. This happens when $x = 0, 2,$ and 4. $f(g(x))$ is also negative for $x > 4$. Alternatively, $f(g(4)) = f(0) = 0$ identifies answer (c).

COMMENT:

You could have students give specific places in the other choices where there was a conflict with information given in the graphs of f and g.

ConcepTests and Answers and Comments for Section 3.5

1. List in order (from smallest to largest) the following functions as regards to the maximum possible value of their slope.

    (a) $\sin x$ (b) $\sin(2x)$ (c) $\sin(3x)$ (d) $\sin(x/2)$

    ANSWER:

    (d), (a), (b), (c). The derivative of $\sin(\alpha x)$ is $\alpha \cos(\alpha x)$. The largest value of $\cos(\alpha x)$ is 1, so the ranking is based only on the value of α.

    COMMENT:

    You could mimic this question replacing the sine function by cosine.

2. At which of the following values of x does $\sin(4\pi x^2)$ attain the largest slope?

    (a) $x = 0$ (b) $x = 1/2$ (c) $x = 1$ (d) $x = 2$

    ANSWER:

    (d). The derivative of $\sin(4\pi x^2)$ is $8\pi x \cos(4\pi x^2)$. At these specific values of x, we have the values $0, -4\pi, 8\pi$, and 16π.

    COMMENT:

    You could mimic this question replacing the sine function by cosine.

3. Which of the graphs are those of $\sin(2x)$ and $\sin(3x)$?

    (a) (I) $= \sin(2x)$, (II) $= \sin(3x)$ (b) (I) $= \sin(2x)$, (III) $= \sin(3x)$
    (c) (II) $= \sin(2x)$, (III) $= \sin(3x)$ (d) (III) $= \sin(2x)$, (IV) $= \sin(3x)$

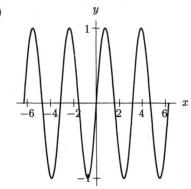

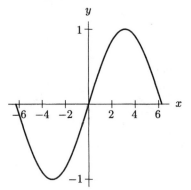

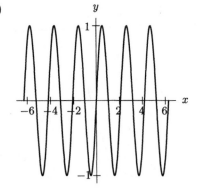

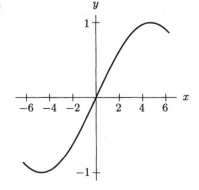

    ANSWER:

    (b). Because $\dfrac{d}{dx} \sin(\alpha x) = \alpha \cos(\alpha x)$, the graph of $\sin(3x)$ will have steeper slopes than $\sin(2x)$. The first positive zeros are at $x = \pi/2$ for $\sin(2x)$ and $x = \pi/3$ for $\sin(3x)$.

    COMMENT:

    Since this is the third time this question has appeared, you may want to emphasize that there are many ways of solving this problem, for example, finding the x-intercepts, remembering the properties of trigonometric functions, and using calculus to look at the slope of the function at various points. You could ask for equations for the graphs in choices (II) and (IV).

4. Which of the graphs is that of $\sin(x^2)$?

(a)

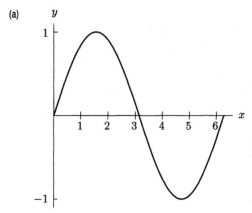

(b)

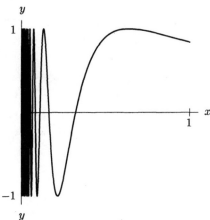

(c)

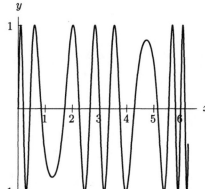

(d)

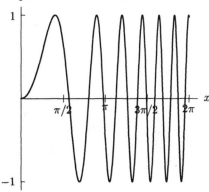

ANSWER:

(d). Since $\dfrac{d}{dx}\sin(x^2) = 2x\cos(x^2)$, the maximum value of the slope of the graph increases as x increases. Also, the zeros of $\sin(x^2)$ are $x = \sqrt{n\pi}$ for $n = 0, 1, 2, 3, \ldots$, which are closer together as x increases.

COMMENT:

You could have students give specific points on the graphs in the other choices which have properties not consistent with those of $y = \sin(x^2)$.

ConcepTests and Answers and Comments for Section 3.8

For Problems 1–2, consider the graphs of $f(x) = (1/2)e^x$ and $g(x) = (1/2)e^{-x}$ in Figure 3.13.

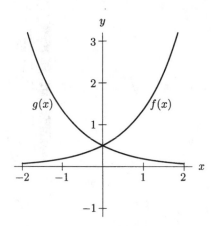

Figure 3.13

1. Which of the following functions represents $f(x) + g(x)$?

(a)

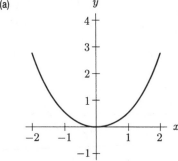

(b)

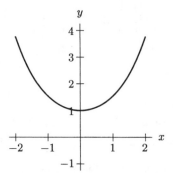

(c)

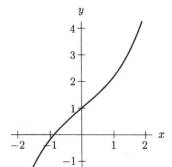

(d)

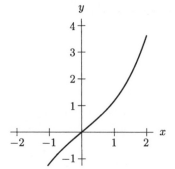

ANSWER:
(b)
COMMENT:
This is one way to introduce $\cosh x$.

2. Which of the following functions represents $f(x) - g(x)$?

(a)

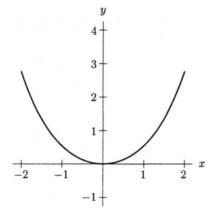

(b)

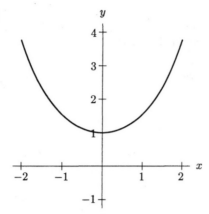

(c)

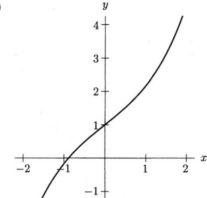

(d)

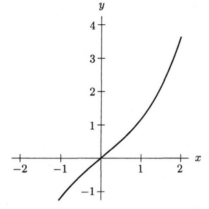

ANSWER:
(d)
COMMENT:
This is one way to introduce $\sinh x$.

ConcepTests and Answers and Comments for Section 3.9

1. In which of the following graphs will using local linearity to approximate the value of the function near $x = c$ give the least error as Δx becomes larger?

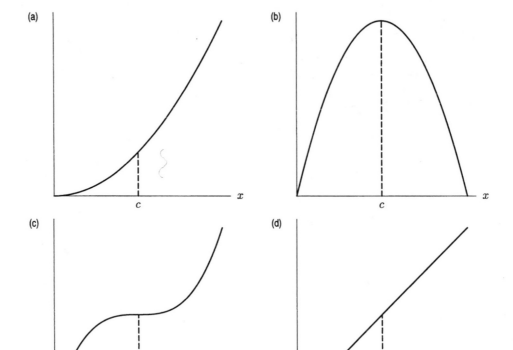

ANSWER:

(d). Local linearity will give exact answers in graph (d) because the tangent line at $x = c$ is identical with the graph on the interval shown.

COMMENT:

You could lay a pencil on the overhead to illustrate the tangent line in each case.

2. To estimate the numerical value of the square root of a number, we use a tangent line approximation about $x = a$. From Figure 3.14, the graph of $\sqrt{x}$, decide for which number the error in using this approximation has the smallest magnitude.

(a) $\sqrt{4.2}$ about $x = 4$
(b) $\sqrt{4.5}$ about $x = 4$
(c) $\sqrt{9.2}$ about $x = 9$
(d) $\sqrt{9.5}$ about $x = 9$
(e) $\sqrt{16.2}$ about $x = 16$
(f) $\sqrt{16.5}$ about $x = 16$

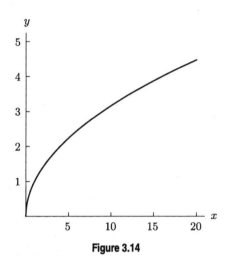

Figure 3.14

ANSWER:

(e). The least error occurs at the point closest to the perfect square whose tangent line has the flattest slope. Thus the answer is (e).

COMMENT:

You could ask the students why the approximation is taken about perfect squares. You could also have students graph $y = \sqrt{x}$ on the window $0 < x < 17, 0 < y < 4.5$ and have their graphing calculator draw tangent lines at the appropriate points.

ConcepTests and Answers and Comments for Section 3.10

1. A function intersects the x-axis at points a and b, where $a < b$. The slope of the function at a is positive and the slope at b is negative. Which of the following is true for any such function? There exists some point on the interval (a, b) where

(a) The slope is zero and the function has a local maximum.
(b) The slope is zero but there is not local maximum.
(c) There is a local maximum but there does not have to be a point at which the slope is zero.
(d) None of the above have to be true.

ANSWER:

(d). The function $f(x) = 1/x^2 - 1$, where $a = -1$ and $b = 1$ is an example where none of the choices, (a), (b), or (c) are valid.

COMMENT:

Rather than look for an equation, you might ask for an example in the form of a graph. You might also ask if requiring the function to be bounded would change their answer.

2. A function intersects the x-axis at points a and b, where $a < b$. The slope of the function at a is positive and the slope at b is negative. Which of the following is true for any such function for which the limit of the function exists at every point? There exists some point on the interval (a, b) where

(a) The slope is zero and the function has a local maximum.

(b) The slope is zero but there is not local maximum.

(c) There is a local maximum but there does not have to be a point at which the slope is zero.

(d) None of the above have to be true.

ANSWER:

(d). The graph in Figure 3.15 is an example of a function for which none of the choices (a), (b), or (c) are valid.

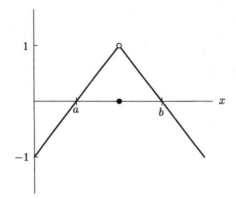

Figure 3.15

COMMENT:

If you graph the function given in the Answer, you might ask students for its equation.

3. A continuous function intersects the x-axis at points a and b, where $a < b$. The slope of the function at a is positive and the slope at b is negative. Which of the following is true for any such function? There exists some point on the interval (a, b) where

(a) The slope is zero and the function has a local maximum.

(b) The slope is zero but there is not local maximum.

(c) There is a local maximum but there does not have to be a point at which the slope is zero.

(d) None of the above have to be true.

ANSWER:

(c). The Extreme Value Theorem guarantees there will be a maximum. The function

$f(x) = (b - a)/2 - |x - (a + b)/2|$ shows that the slope of such a function need not be zero.

COMMENT:

This answer is more believable if the function is graphed.

4. A continuous and differentiable function intersects the x-axis at points a and b, $(a < b)$. The slope of the function at a is positive and the slope at b is negative. Which of the following is true for any such function? There exists some point on the interval (a, b) where

(a) The slope is zero and the function has a local maximum.

(b) The slope is zero but there is not local maximum.

(c) There is a local maximum but there does not have to be a point at which the slope is zero.

(d) None of the above have to be true.

ANSWER:

(a). The Extreme Value Theorem guarantees there will be a maximum, while the Mean Value Theorem requires a horizontal tangent on the interval.

COMMENT:

You might review the hypothesis of the Mean Value Theorem.

Chapter Four

ConcepTests and Answers and Comments for Section 4.1 ⎯⎯⎯⎯⎯⎯

For Problems 1–3, consider the graph of $f(x)$ in Figure 4.1.

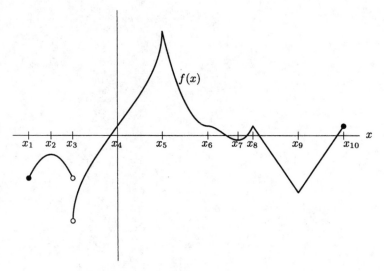

Figure 4.1

1. How many critical points does f have?
   ANSWER:
   There are 6 critical points. One occurring at x_2, x_5, x_6, x_7, x_8, and x_9.
   COMMENT:
   You might note that endpoints are not defined to be critical points and critical points can only occur where the function exists.

2. How many local minima does f have?
   ANSWER:
   There are 3 local minima. They occur at x_1, x_7, and x_9.
   COMMENT:
   Follow-up Question. What happens if one of the holes at x_3 is filled in? Does it matter which hole?

3. Consider the graph of $y = f(x)$ in Figure 4.1. How many local maxima does f have?
   ANSWER:
   There are 4 local maxima. They occur at x_2, x_5, x_8, and x_{10}.
   COMMENT:
   Follow-up Question. What happens if one of the holes at x_3 is filled in? Does it matter which hole?

4. Concerning the graph of the function in Figure 4.2, which of the following statements is true?

(a) The derivative is zero at two values of x, both being local maxima.

(b) The derivative is zero at two values of x, one is a local maximum while the other is a local minimum.

(c) The derivative is zero at two values of x, one is a local maximum on the interval while the other is neither a local maximum nor a minimum.

(d) The derivative is zero at two values of x, one is a local minimum on the interval while the other is neither a local maximum nor a minimum.

(e) The derivative is zero only at one value of x where it is a local minimum.

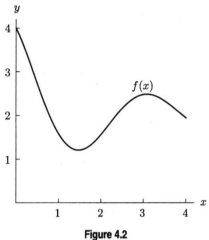

Figure 4.2

ANSWER:

(b). The derivative is zero where it has a horizontal tangent, having a local maximum if it is concave down there, a local minimum if it is concave up.

COMMENT:

You could sketch a different graph and ask the same question.

5. If the graph in Figure 4.3 is that of $f'(x)$, which of the following statements is true concerning the function f?

(a) The derivative is zero at two values of x, both being local maxima.

(b) The derivative is zero at two values of x, one is a local maximum while the other is a local minimum.

(c) The derivative is zero at two values of x, one is a local maximum on the interval while the other is neither a local maximum nor a minimum.

(d) The derivative is zero at two values of x, one is a local minimum on the interval while the other is neither a local maximum nor a minimum.

(e) The derivative is zero only at one value of x where it is a local minimum.

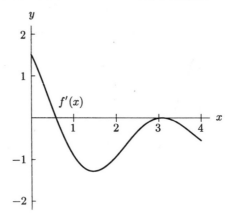

Figure 4.3

ANSWER:

(c). When $x \approx 0.6$, the derivative is positive to the left and negative to the right of that point. This gives a local maximum. When $x \approx 3.1$, the graph of the derivative is below the axis (negative) on both sides of this point. This is neither a local minimum nor a maximum.

COMMENT:

You could sketch a different graph and ask the same question.

ConcepTests and Answers and Comments for Section 4.2

1. The functions in Figure 4.4 have the form $y = A \sin x$. Which of the functions has the largest A? Assume the scale on the vertical axes is the same for each graph.

(a)
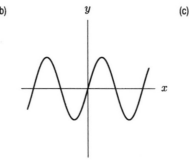

(b)

(c)
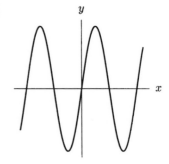

Figure 4.4

ANSWER:

(c) has the largest amplitude.

COMMENT:

Remind students that they only need to compare the graphs. They do not need any values on the axes labeled, but they need to know that the vertical scales are the same on all three graphs.

2. The graphs in Figure 4.5 have the form $y = \sin Bx$. Which of the functions has the largest B? Assume the scale on the horizontal axes is the same for each graph.

(a)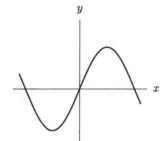

(b)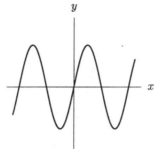

(c)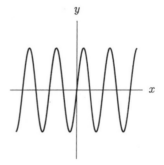

Figure 4.5

ANSWER:

(c). Since period $= 2\pi/|B|$, then the largest B corresponds to the smallest period.

COMMENT:

Remind the students that the horizontal scale is the same for all three graphs.

3. Let $f(x) = ax + b/x$. What are the critical points of $f(x)$?

(a) $-b/a$ (b) 0 (c) $\pm\sqrt{b/a}$

(d) $\pm\sqrt{-b/a}$ (e) No critical points

ANSWER:

(c) or (e). $f'(x) = a - b/x^2$. Therefore the critical points, if they exist, of $f(x)$ are $x = \pm\sqrt{b/a}$. Thus the answer is (c) when a, b are either both positive or both negative, (e) when a and b have opposite signs.

COMMENT:

Students have a hard time with parameters. To get them warmed up, you may ask the same question where you have assigned specific values to a and b. Note that the next two questions follow-up on this one.

4. Let $f(x) = ax + b/x$. Suppose a and b are positive. What happens to $f(x)$ as b increases?

(a) The critical points move further apart.
(b) The critical points move closer together.
(c) The critical values move further apart.
(d) The critical values move closer together.

ANSWER:

(a) and (c). Since the critical points of $f(x)$ are $x = \pm\sqrt{b/a}$, then as b increases $\left|\pm\sqrt{b/a}\right|$ increases. Therefore (a).

The critical values of $f(x)$ are $f(\pm\sqrt{b/a}) = a(\pm\sqrt{b/a}) + \dfrac{b}{\pm\sqrt{b/a}} = \pm\sqrt{ab} \pm \sqrt{ab} = \pm2\sqrt{ab}$. Therefore (c) is also true.

COMMENT:

You may also ask what happens if a and b are negative, or what happens if a and b have different signs.

5. Let $f(x) = ax + b/x$. Suppose a and b are positive. What happens to $f(x)$ as a increases?

(a) The critical points move further apart.
(b) The critical points move closer together.
(c) The critical values move further apart.
(d) The critical values move closer together.

ANSWER:

(b) and (c). Since the critical points of $f(x)$ are $x = \pm\sqrt{b/a}$, then as a increases $\left|\pm\sqrt{b/a}\right|$ decreases. Therefore (b). The critical values of $f(x)$ are $f(\pm\sqrt{b/a}) = a(\pm\sqrt{b/a}) + \dfrac{b}{\pm\sqrt{b/a}} = \pm\sqrt{ab} \pm \sqrt{ab} = \pm2\sqrt{ab}$. Therefore (c) is also true.

COMMENT:

You might also ask what happens if a and b are negative, or what happens if a and b have different signs.

ConcepTests and Answers and Comments for Section 4.3

For Problems 1–2, consider the graph of $f(x)$ in Figure 4.6.

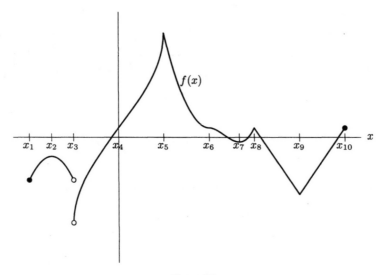

Figure 4.6

1. Where is the global maximum of $f(x)$?

    ANSWER:

    x_5

    COMMENT:

    Follow-up Question. What happens to the global maximum if f is defined only on the open interval from x_1 to x_4?
    Answer. There is no global maximum.

2. Where is the global minimum of $f(x)$?

    ANSWER:

    There is no global minimum.

    COMMENT:

    Follow-up Question. What happens to the global minimum if f is defined only on the closed interval from x_5 to x_{10}?

    Answer. x_9 becomes the global minimum.

3. A global maximum is always a critical point.

    (a) True
    (b) False

    ANSWER:

    (b). If a global maximum occurs at the endpoint of the interval, it is not a critical point.
    COMMENT:

    This question should remind the student of the definition of a critical point.

4. A function defined on all points of a closed interval always has a global maximum *and* a global minimum.

    (a) True
    (b) False

    ANSWER:

    (b). The function has to be continuous for this to always be true.
    COMMENT:

    Have students give counterexamples.

ConcepTests and Answers and Comments for Section 4.4

1. Which is correct? A company should

   (a) Maximize revenue
   (b) Maximize marginal revenue
   (c) Minimize cost
   (d) Minimize marginal cost
   (e) None of the above

   ANSWER:

   (e). A company should maximize profit, that is, it should maximize revenue minus cost.
   COMMENT:
   Follow-up Question. Where are (a) and (c) incorrect?
   Answer. Answer (a) is incorrect because revenue always increases as the quantity of goods increases, so there is no maximum. Answer (c) is incorrect because the minimum cost occurs when the company produces zero goods.

2. Which is correct? A company can always increase its profit by increasing production if, at its current level of production,

   (a) Marginal revenue − Marginal cost > 0
   (b) Marginal revenue − Marginal cost $= 0$
   (c) Marginal revenue − Marginal cost < 0
   (d) Marginal revenue − Marginal cost is increasing

   ANSWER:

   (a)
   COMMENT:
   Have the students discuss each of the scenarios (b)–(d)? if the company increases production.

ConcepTests and Answers and Comments for Section 4.5

1. Given that $f'(x)$ is continuous everywhere and changes from negative to positive at $x = a$, which of the following statements must be true?

   (a) a is a critical point of $f(x)$
   (b) $f(a)$ is a local maximum
   (c) $f(a)$ is a local minimum
   (d) $f'(a)$ is a local maximum
   (e) $f'(a)$ is a local minimum

   ANSWER:

   (a) and (c). a is a critical point and $f(a)$ is a local minimum.
   COMMENT:
   Follow-up Question. What additional information would you need to determine whether $f(a)$ is also a global maximum?
   Answer. You need to know the values of the function at all local minima.

2. On the same side of a straight river are two towns, and the townspeople want to build a pumping station, S. See Figure 4.7. The pumping station is to be at the river's edge with pipes extending straight to the two towns. Which function must you minimize over the interval $0 \leq x \leq 4$ to find the location for the pumping station that minimizes the total length of the pipe?

(a) $1 + |x| + |4 - x| + 4$
(b) $\sqrt{x^2 + 1} + \sqrt{(4 - x)^2 + 16}$
(c) $(1/2) \cdot x \cdot 1 + (1/2) \cdot (4 - x) \cdot 4$
(d) $x(4 - x)$

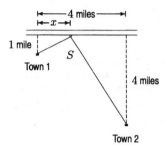

Figure 4.7

ANSWER:

(b). By the Pythagorean Theorem,

$$\text{Distance from Town 1 to } S = \sqrt{x^2 + 1^2}$$
$$\text{Distance from Town 2 to } S = \sqrt{(4 - x)^2 + 4^2},$$

so the sum of these distances is given by (b).

COMMENT:

Follow-up Question. What function must be minimized if construction of the pipeline from Town 1 to the river is twice as expensive per foot as construction of the pipeline from Town 2 to the river and the goal is to minimize total construction cost?

Answer. $2\sqrt{x^2 + 1} + \sqrt{(4 - x)^2 + 16}$.

ConcepTests and Answers and Comments for Section 4.6

1. A spherical snowball of radius r cm has surface area S cm^2. As the snowball gathers snow, its radius increases as in Figure 4.8. Approximately how fast, in cm^2/min, is S increasing when the radius is 20 cm?

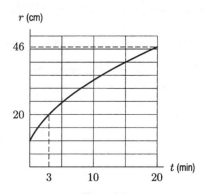

Figure 4.8

(a) $4\pi \cdot 3^2$

(b) $4\pi \cdot 20^2$

(c) $4\pi \cdot 46^2$

(d) $8\pi \cdot 3 \cdot 2.5$

(e) $8\pi \cdot 20 \cdot 2.5$

(f) $8\pi \cdot 46 \cdot 1$

ANSWER:

(e). We want dS/dt when $r = 20$, that is, at $t = 3$. The surface area is given by

$$S = 4\pi r^2,$$

so differentiating gives

$$\frac{dS}{dt} = 8\pi r \frac{dr}{dt}.$$

To find dr/dt, we draw a tangent line to the curve at $t = 3$ and estimate its slope to be 2.5 cm/min. Thus

$$\frac{dS}{dt} = 8\pi \cdot 20 \cdot 2.5 \text{ cm}^2/\text{min}.$$

COMMENT:

Have students check the units of their answer.

2. A car is driving along a straight flat road when a plane flies overhead. Let x miles be the distance traveled by the car and y miles be the distance traveled by the plane at time t in hours since the plane was directly over the car. At time t, the distance D miles between the car and the plane is given by

$$D^2 = x^2 + y^2 + 2^2.$$

At one moment, the car has gone 3 miles and is moving at 60 mph, and the plane has gone 30 miles and is moving at 500 mph. To find the rate at which D is increasing at that time, you should:

(a) Substitute $x = 3$, $y = 30$ into $D^2 = x^2 + y^2 + 2^2$.

(b) Differentiate $D^2 = x^2 + y^2 + 2^2$ after substituting $x = 3$, $y = 30$. Then substitute $dx/dt = 60$, $dy/dt = 500$.

(c) Differentiate $D^2 = x^2 + y^2 + 2^2$. Then substitute $x = 3$, $y = 30$, $dx/dt = 60$, $dy/dt = 500$.

(d) None of the above.

ANSWER:

(c). Note that substituting $x = 3$, $y = 30$ before differentiating, as in (b), gives zero instead of the correct answer.

COMMENT:

Have the students calculate the answer, which is $15{,}180/\sqrt{913} = 502.385$ mph.

3. Let $P(t)$ be the population of California in year t. Then $P'(2005)$ represents:

   (a) The growth rate (in people per year) of the population.
   (b) The growth rate (in percent per year) of the population.
   (c) The approximate number of people by which the population increased in 2005.
   (d) The approximate percent increase in the population in 2005.
   (e) The average yearly rate of change in the population since $t = 0$.
   (f) The average yearly percent rate of change in the population since $t = 0$.

   ANSWER:

   (a). The derivative represents the rate of change of $P(t)$ in people per year. Answer (c) is approximately correct.
   COMMENT:

   Using P, have students write expressions for each of the other rates.

4. Let $g(t)$ be the average number of email messages, in millions per day, sent in a particular region in year t. What are the units and interpretations of the following quantities?

   (a) $g(2005)$
   (b) $g'(2005)$
   (c) $g^{-1}(20)$
   (d) $(g^{-1})'(20)$

   ANSWER:

   (a) The average number of email messages sent per day in 2005; the units are millions of messages per day.
   (b) The rate of change in the number of messages per day; the units are millions of messages per day per year.
   (c) The year in which 20 million messages are sent per day; the units are years.
   (d) The derivative is the rate of change of the year with respect to the number of messages per day. To interpret this derivative, think of it as the approximate time in years it takes for the email traffic to increase by 1 million per day when it is already 20 million per day. The units are years per million messages per day.

   COMMENT:

   Additional practice can be provided by asking students to convert verbal descriptions of similar quantities to symbols.

5. The light in the lighthouse in Figure 4.9 rotates at 2 revolutions per minute. To calculate the speed at which the spot of light moves along the shore, it is best to differentiate:

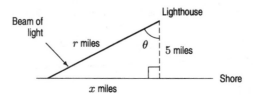

Figure 4.9

   (a) $r^2 = 5^2 + x^2$
   (b) $x = r \sin \theta$
   (c) $x = 5 \tan \theta$
   (d) $r^2 = 2^2 + x^2$

   ANSWER:
   (c). Differentiating $x = 5 \tan \theta$ gives

   $$\frac{dx}{dt} = \frac{5}{\cos^2 \theta} \frac{d\theta}{dt}.$$

Since the light rotates at 2 revolutions per minute = 4π radians per minute, we know $d\theta/dt = 4\pi$. Thus we can calculate dx/dt, the speed at which the spot is moving, for any angle θ.

   Differentiating any of the other relationships introduces dr/dt, whose values we cannot find as easily as we can find $d\theta/dt$.

6. The foot of the ladder in Figure 4.10 moves away from the wall at a speed of 2 ft/min, causing the top of the ladder to slide down the wall without leaving it. Label each of the following statements as True or False and give a reason.

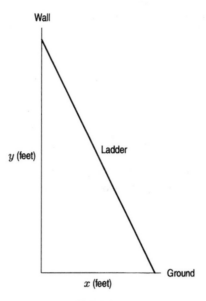

Figure 4.10

(a) dx/dt and dy/dt have the same sign.
(b) The top of the ladder is moving faster and faster.
(c) Keeping dx/dt constant, doubling x, y, and the length of the ladder, doubles dy/dt.

ANSWER:

We know $dx/dt = 2$ ft/min. Let k be the (constant) length of the ladder in feet, so

$$x^2 + y^2 = k^2.$$

Differentiating with respect to t gives

$$2x\frac{dx}{dt} + 2y\frac{dy}{dt} = 0$$

$$\frac{dy}{dt} = -\frac{x}{y}\frac{dx}{dt} = -2\frac{x}{y}.$$

(a) False, since $dx/dt > 0$ and $dy/dt < 0$. Alternatively, this could be done without computation. Because the top of the ladder moves down, y decreases, so dy/dt is negative. We are given $dx/dt = 2$, which is positive, so dx/dt and dy/dt have different signs.
(b) True, since as x increases, y decreases, so x/y increases. Thus, the magnitude of dy/dt increases, which means that the top of the ladder is moving faster and faster.
(c) False, since doubling x and y leaves the ratio x/y unchanged. Thus, dy/dt is unchanged.

COMMENT:

Have students carry out some of the calculation with specific values of x and y to confirm their answers.

ConcepTests and Answers and Comments for Section 4.7

1. Which of the limits cannot be computed with L'Hospital's rule?

   (a) $\lim\limits_{x \to 0} \dfrac{\sin x}{x}$

   (b) $\lim\limits_{x \to 0} \dfrac{\cos x}{x}$

   (c) $\lim\limits_{x \to 0} \dfrac{x}{\sin x}$

   (d) $\lim\limits_{x \to \infty} \dfrac{x}{e^x}$

   ANSWER:

   (b)

   COMMENT:

   What is the limit in each case? (The limit in (b) does not exist.)

2. Arrange in order by dominance as $x \to \infty$, from least to most dominant.

   (a) x^{100}

   (b) $x(\ln x)^2$

   (c) $\dfrac{e^{2x}}{x}$

   (d) x^2

   (e) $e^x \ln x$

   ANSWER:

   (b), (d), (a), (e), (c). As $x \to \infty$, in increasing order of of dominance, we have $x(\ln x)^2 < x^2 < x^{100} < e^x \ln x < e^{2x}/x$.

   COMMENT:

   Do the students know the answer without using l'Hopital's Rule?

3. For which of the following can you use l'Hopital's rule to evaluate the limit?

   (a) $\lim\limits_{x \to 0} \dfrac{\cos x}{x}$

   (b) $\lim\limits_{x \to 0} \dfrac{\sin x}{x}$

   (c) $\lim\limits_{x \to \infty} e^{-x} x^2$

   (d) $\lim\limits_{x \to 0} e^{-x} x^2$

   ANSWER:

   (b), (c) after rewriting it as $\lim\limits_{x \to \infty} \dfrac{x^2}{e^x}$.

   COMMENT:

   Follow-up Question. Compute the limits.

   Answer. For (a) the limit does not exist; for (b) the limit is 1; for (c) and (d) the limit is zero.

4. True or false?

   (a) Suppose $f(a) = g(a) = 0$. According to l'Hopital's rule, $\dfrac{f(a)}{g(a)} = \dfrac{f'(a)}{g'(a)}$.

   (b) Suppose $f(a) = g(a) = 0$. Let $Q(x) = f(x)/g(x)$. Then $\lim\limits_{x \to a} Q(x) = \lim\limits_{x \to a} Q'(x)$.

   (c) Suppose $f(a) = 0$ and $\lim\limits_{x \to a} g(x) = \infty$. Then $\lim_{x \to a} f(x)g(x) = \lim_{x \to a} f'(x)g'(x)$.

   ANSWER:

   (a) False. l'Hopital's rule concerns the equality of limits, not the equality of functions.

   (b) False. In applying l'Hopital's rule, the numerator and the denominator are differentiated separately.

   (c) False. To apply l'Hopital's rule to a product, the expression must first be rewritten as a ratio.

   COMMENT:

   Follow-up Question. What does l'Hopital's rule say?

   Answer. If f and g are differentiable, $f(a) = g(a) = 0$, and $g'(a) \neq 0$, then

   $$\lim\limits_{x \to a} \frac{f(x)}{g(x)} = \frac{f'(a)}{g'(a)}.$$

ConcepTests and Answers and Comments for Section 4.8

1. Match the parameterizations (a)–(e) and curves I–V.

(a) $x = \sin t$ $y = -t$
(b) $x = |3t|$ $y = t$
(c) $x = \sin t$ $y = \cos t$
(d) $x = 3t^3$ $y = t^3$
(e) $x = |3t|$ $y = 3t$

I.   Line
II. Circle
III. Right angle
IV. Acute angle
V.   Wave

ANSWER:
(a)—V, (b)—IV, (c)—II, (d)—I, (e)—III
COMMENT:
If your students are using graphing calculators to answer this question, then they will need to make sure that the range of t is both positive and negative.

2. The equation of the line tangent to the curve $x = \sin 3t$, $y = \sin t + 1$ at the point where $t = 0$ is

(a) $y = 3x$
(b) $y = 3x + 1$
(c) $y = (1/3)x$
(d) $y = (1/3)x + 1$
(e) none of the above

ANSWER:
(d). Since $\dfrac{dy}{dx} = \dfrac{dy/dt}{dx/dt}$, the slope of the tangent line at $t = 0$ is

$$\left.\frac{dy}{dx}\right|_{x=0} = \left.\frac{\cos t}{3\cos 3t}\right|_{t=0} = \frac{1}{3}.$$

The line goes through the point $(0, 1)$, so the equation is $y - 1 = (1/3)x$.
COMMENT:
You could have students graph the curve and find the equation of the tangent line at other points, for example where $t = \pi/2$.

Chapter Five

ConcepTests and Answers and Comments for Section 5.1

1. Consider the graph in Figure 5.1.

   (a) Give an interval on which the left-hand sum approximation of the area under the curve on that interval is an underestimate.

   (b) Give an interval on which the left-hand sum approximation of the area under the curve on that interval is an overestimate.

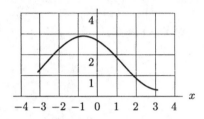

Figure 5.1

ANSWER:

(a) $[-3, -1]$ because the function is increasing there.

(b) $[0, 3]$ because the function is decreasing there.

COMMENT:

Follow-up Question. What about right-hand sum?

Answer. On the interval $[0, 3]$ the right-hand sum is an underestimate because the function is decreasing there. On the interval $[-3, -1]$ the right-hand sum is an overestimate because the function is increasing there.

2. The velocities of two cyclists, traveling in the same direction, are given in Figure 5.2. If initially the two cyclists are alongside each other, when does Cyclist 2 overtake Cyclist 1?

   (a) Between 0.75 and 1.25 minutes

   (b) Between 1.25 and 1.75 minutes

   (c) Between 1.75 and 2.25 minutes

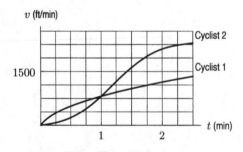

Figure 5.2

ANSWER:

(b). Between 1.25 and 1.75 minutes, because the area under the two curves is about equal at some point in this interval.

COMMENT:

You could ask the students to answer the question if either, or both, of the statements "traveling in the same direction" or "If initially the two cyclists were alongside each other" are removed. You could also use this question to review the concepts of derivative.

Follow-up Question. When the cyclists pass each other, which one is accelerating more?

Answer. Cyclist 2 is accelerating more because between 1.25 and 1.75 minutes the slope of the velocity graph at every point in that interval is steeper than the slope corresponding to Cyclist 1.

3. The table gives a car's velocity at time t. In each of the four fifteen-minute intervals, the car is either always speeding up or always slowing down. The car's route is a straight line with four towns on it. Town A is 60 miles from the starting point, town B is 70 miles from the starting point, town C is 73 miles from the starting point, and town D is 80 miles from the starting point. For each town, decide if the car has passed it, has not yet passed it, or may have passed it.

t (minutes)	0	15	30	45	60
v (miles per hour)	60	75	72	78	65

ANSWER:
Calculating an upper and a lower estimate for the distance traveled, we discover that the car traveled at least 67.25 miles and at most 76.5 miles:

$$\text{Lower estimate } = (60 + 72 + 72 + 65)0.25 = 67.25$$
$$\text{Upper estimate } = (75 + 75 + 78 + 78)0.25 = 76.50.$$

Thus, the car definitely passed town A, may have passed towns B and C, and definitely did not pass town D.
COMMENT:
Note that the velocity of the car is not monotone. Therefore the upper and lower estimates for the distance traveled are not equal to the right and left sum estimates.

ConcepTests and Answers and Comments for Section 5.2

1. Figure 5.3 shows the graph of $y = x^2$. The right-hand sum for eight equal divisions is given by
   (a) $0.5^2 + 1^2 + 1.5^2 + 2^2 + 2.5^2 + 3^2 + 3.5^2 + 4^2$
   (b) $0.5(0.5) + 1(0.5) + 1.5(0.5) + 2(0.5) + 2.5(0.5) + 3(0.5) + 3.5(0.5) + 4(0.5)$
   (c) $0.5^2(0.5) + 1^2(0.5) + 1.5^2(0.5) + 2^2(0.5) + 2.5^2(0.5) + 3^2(0.5) + 3.5^2(0.5) + 4^2(0.5)$
   (d) $0^2(0.5) + 0.5^2(0.5) + 1^2(0.5) + 1.5^2(0.5) + 2^2(0.5) + 2.5^2(0.5) + 3^2(0.5) + 3.5^2(0.5)$

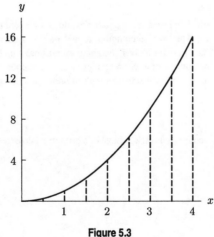

Figure 5.3

ANSWER:
(c). The width of each rectangle is 0.5, and the x values are squared.
COMMENT:
You could ask the students to give the left-hand sum also. You could also comment on how the right-hand sum gives an overestimate because the function is increasing.

2. For the following question consider continuous curves that are increasing and concave up. As the number of rectangles (of equal width) triples in Figures 5.4 and 5.5, the areas represented by $A1$ and $A2$ decrease.

(a) True
(b) False
(c) It depends on the graph

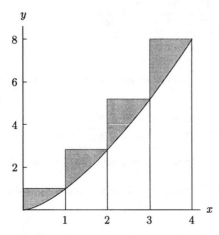

Figure 5.4: $A1 = Shaded\ Region$

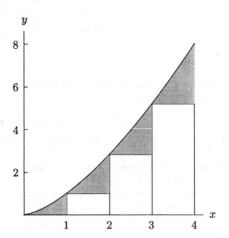

Figure 5.5: $A2 = Shaded\ Region$

ANSWER:

(a). In both graphs, taking more rectangles of equal width will decrease the values of $A1$ and $A2$, with the area of the rectangles in both figures becoming closer to the area under the curve.

COMMENT:

You could hand out the graphs to your students and have them subdivide and conjecture. You could ask the students the same question, but without the constraint that the function be increasing and concave up.

3. As the number of rectangles (of equal width) in Figures 5.4 and 5.5 goes to infinity which of the following statements is true? Consider only continuous curves.

(a) $A1$ goes to zero but $A2$ does not. The lower rectangular area does not equal the upper rectangular area.
(b) $A2$ goes to zero but $A1$ does not. The lower rectangular area does not equal the upper rectangular area.
(c) $A2$ and $A1$ don't go to zero. The upper and lower rectangular areas are equal.
(d) $A1$ and $A2$ both go to zero. The upper and lower rectangular areas are equal.
(e) $A1$ and $A2$ both go to zero. The upper and lower rectangular areas are not equal.

ANSWER:

(d)

COMMENT:

Students should realize that all continuous curves have this property. Have them explore this idea by drawing arbitrary continuous curves.

ConcepTests and Answers and Comments for Section 5.3

1. A bicyclist starts from home and rides back and forth along a straight east/west highway. Her velocity is given in Figure 5.6 (positive velocities indicate travel toward the east, negative toward the west).

   (a) On what time intervals is she stopped?
   (b) How far from home is she the first time she stops, and in what direction?
   (c) At what time does she bike past her house?
   (d) If she maintains her velocity at $t = 11$, how long will it take her to get back home?

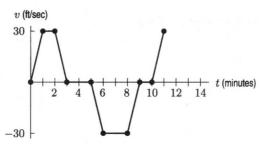

Figure 5.6

ANSWER:

(a) On $[3, 5]$ and on $[9, 10]$, since $v = 0$ there.
(b) 3600 feet to the east, since this is the area under the velocity curve between $t = 0$ and $t = 3$.
(c) At $t = 8$ minutes, since the areas above and below the curve between $t = 0$ and $t = 8$ are equal.
(d) It will take her 30 seconds longer. By calculating areas, we see that at $t = 11$,

$$\text{Distance from home } = 2 \cdot 30 \cdot 60 - 3 \cdot 30 \cdot 60 + 0.5 \cdot 30 \cdot 60 = -900 \text{ feet.}$$

Thus, at $t = 11$, she is 900 feet west of home. At a velocity of 30 ft/sec eastward, it takes $900/30 = 30$ seconds to get home.

COMMENT:

To get the correct distance we have to convert minutes to seconds.

2. Arrange the intervals in the order for which the average value of $\sin x$ over the interval increases from smallest to largest.

   (a) $0 \leq x \leq \pi$
   (b) $\pi/2 \leq x \leq 3\pi/2$
   (c) $\pi \leq x \leq 2\pi$
   (d) $3.14 \leq x \leq 3.15$

   ANSWER:

   (c), (d), (b), (a). On $0 \leq x \leq \pi$, the average value of $\sin x$ is positive; on $\pi/2 \leq x \leq 3\pi/2$, the average value is 0; on $\pi \leq x \leq 2\pi$, the average value is negative. Since $\sin \pi = 0$ and $\pi = 3.14159\ldots$, the sine function changes sign in the interval $3.14 \leq x \leq 3.15$. Since π is closer to the left-end of $3.14 \leq x \leq 3.15$, and because $\sin x$ changes from positive to negative and the graph of $\sin x$ is symmetric about $x = \pi$, the average value of $\sin x$ on this interval is small and negative.

   COMMENT:

   Follow-up Question. What is the average value of $\sin x$ over each interval?

   Answer. (a) The average value is $\dfrac{1}{\pi} \displaystyle\int_0^\pi \sin x \, dx = 0.6366$. (b) The average value is 0. (c) The average value is

   $\dfrac{1}{\pi} \displaystyle\int_\pi^{2\pi} \sin x \, dx = -0.6366$. (d) The average value is $\dfrac{1}{0.01} \displaystyle\int_{3.14}^{3.15} \sin x \, dx = -3.407 \times 10^{-3}$.

3. Arrange the intervals in the order for which the average value of $\cos x$ over the interval increases from smallest to largest.

   (a) $0 \leq x \leq \pi$
   (b) $\pi/2 \leq x \leq 3\pi/2$
   (c) $4.71 \leq x \leq 4.72$
   (d) $3.14 \leq x \leq 3.15$

   ANSWER:

   (d), (b), (a), (c). On $0 \leq x \leq \pi$, the average value of $\cos x$ is 0; on $\pi/2 \leq x \leq 3\pi/2$, the average value is negative. Since $3\pi/2$ is closer to the left-end of $4.71 \leq x \leq 4.72$, and because $\cos x$ changes from negative to positive, and the graph of $\cos x$ is symmetric about $x = 3\pi/2$, the average value of $\cos x$ on this interval is small and positive. On $3.14 \leq x \leq 3.15$, the average value is almost -1.

   COMMENT:

   This is an excellent question to explore graphically.

ConcepTests and Answers and Comments for Section 5.4

1. Figure 5.7 contains the graph of $F(x)$, while the graphs in (a)–(d) are those of $F'(x)$. Which shaded region represents $F(b) - F(a)$?

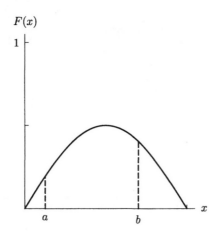

Figure 5.7

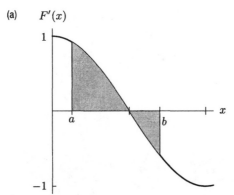

(a)

(b)

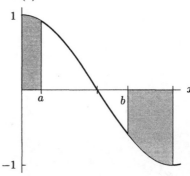

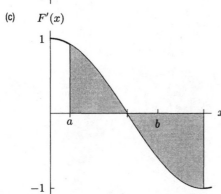

(c)

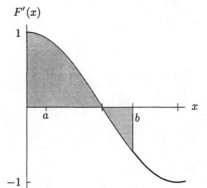

(d)

ANSWER:

(a). Because $F(b) - F(a) = \int_a^b F'(x)dx$ the left-hand and right-hand limits of the integral must be a and b, respectively.

COMMENT:

Follow-up Question. What feature of the graph of $F'(x)$ tells you that $F(b) > F(a)$?

Answer. More area is shaded above the x-axis than below. Thus, $F(b) - F(a) > 0$ so $F(b) > F(a)$.

Chapter Six

ConcepTests and Answers and Comments for Section 6.1

1. Which of the following graphs (a)–(d) could represent an antiderivative of the function shown in Figure 6.1?

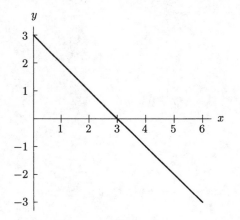

Figure 6.1

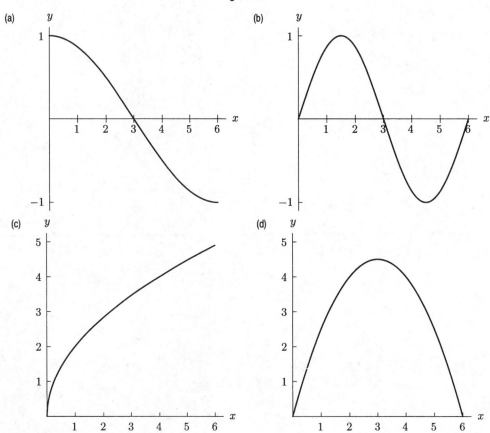

ANSWER:

(d). Because the graph in Figure 6.1 is decreasing, the graph of the antiderivative must be concave down. Because the graph in Figure 6.1 is positive for $x < 3$, zero for $x = 3$, and negative for $x > 3$, then $x = 3$ is a local maximum.

COMMENT:

You can mimic this problem with $y = x - 3$.

2. Which of the following graphs (a)–(d) could represent an antiderivative of the function shown in Figure 6.2?

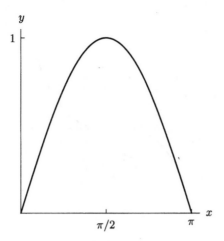

Figure 6.2

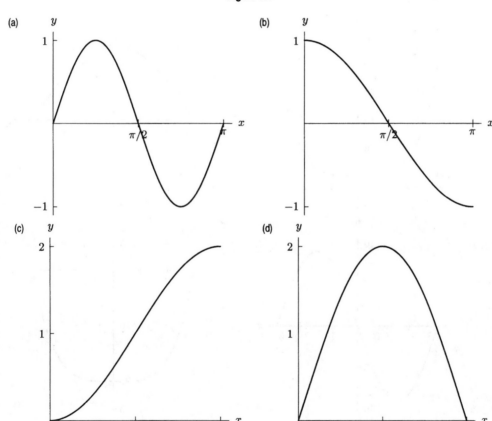

ANSWER:

(c). Because the graph in Figure 6.2 is always positive on this interval, the antiderivative must be increasing for this interval.

COMMENT:

You can mimic this problem with $y = -\sin x$. You could also ask your students to draw other possible antiderivatives of the function shown in Figure 6.2.

3. Which of the following graphs (a)–(d) could represent an antiderivative of the function shown in Figure 6.3?

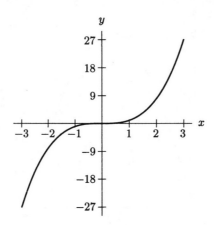

Figure 6.3

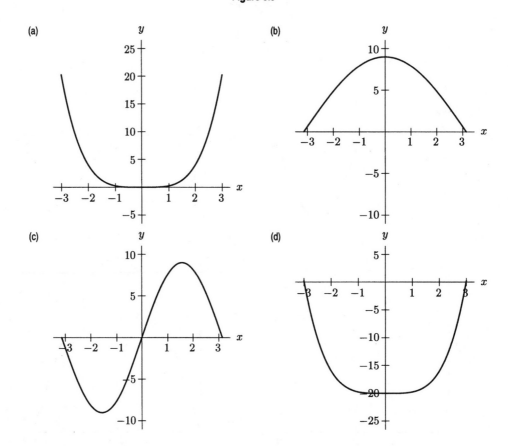

ANSWER:

(a) and (d). Because the graph in Figure 6.3 is continually increasing, the graph of its antiderivative is concave up. Notice that the graphs in (a) and (d) differ only by a vertical shift.

COMMENT:

You could also point out that since the graph in Figure 6.3 is negative from $-3 < x < 0$, then the graph of an antiderivative must be decreasing on this interval. Also, since the graph in Figure 6.3 is positive for $0 < x < 3$, then the graph of an antiderivative must be increasing on this interval.

4. Consider the graph of $f'(x)$ in Figure 6.4. Which of the functions with values from the Table 6.1 could represent $f(x)$?

Table 6.1

	x	0	2	4	6
(a)	$g(x)$	1	3	4	3
(b)	$h(x)$	5	7	8	7
(c)	$j(x)$	32	34	35	34
(d)	$k(x)$	−9	−7	−6	−7

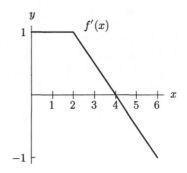

Figure 6.4

ANSWER:

(a), (b), (c), (d)

COMMENT:

You might point out only relative values of functions are important for this problem, not the actual values.

5. Graphs of the *derivatives* of four functions are shown in (I)–(IV). For the *functions* (not the derivative) list in increasing order which has the greatest change in value on the interval shown.

(a) (I), (IV), (III), (II)

(b) (I), (IV), (II), (III)

(c) (I) = (II), (IV), (III)

(d) (I) = (II), (III) = (IV)

(e) (I) = (II) = (III) = (IV)

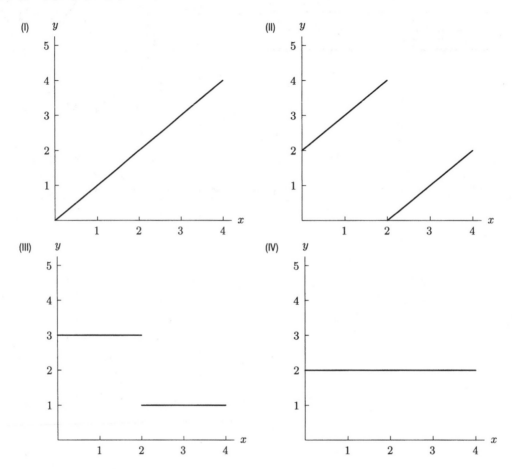

ANSWER:

(e). The ordering will be given by the values of the area under the derivative curves. These areas are (I) $(1/2)(4)(4) = 8$, (II) $(1/2)(2)(4 + 2) + (1/2)(2)(2) = 8$, (III) $2(3) + 2(1) = 8$, and (IV) $4(2) = 8$.

COMMENT:

You can draw some other graphs of derivatives and ask which function has the greatest change over a specified interval.

6. Graphs of the *derivatives* of four functions are shown in (I) - (IV). For the *functions* (not the derivative) list in increasing order which has the greatest change in value on the interval shown.

(a) (I), (III), (IV), (II)

(b) (I) = (III), (IV), (II)

(c) (IV), (I) = (III), (II)

(d) (I) = (III) = (IV), (II)

(e) (I) = (II) = (III) = (IV)

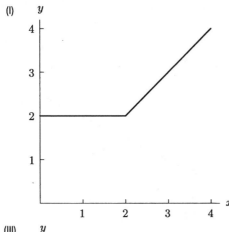

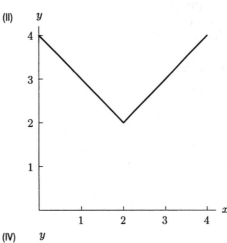

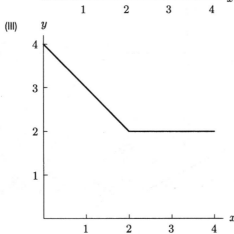

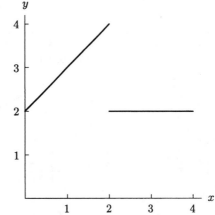

ANSWER:

(d). The ordering will be given by the values of the area under these curves. These areas are (I) $4(2) + (1/2)(2)(2) = 10$, (II) $4(2) + 2(1/2)(2)(2) = 12$, (III) same as (I), 10, and (IV) $(1/2)(2)(2 + 4) + 2(2) = 10$.

COMMENT:

You could ask students how to change graph (II) to have area equal to 10 and not be identical to any of the other graphs.

ConcepTests and Answers and Comments for Section 6.3 ━━━━━━━━

1. Graphs (I)–(III) show velocity versus time for three different objects. Order graphs (I)–(III) in terms of the distance traveled in four seconds. (Greatest to least)

   (a) (I), (II), (III)
   (b) (III), (II), (I)
   (c) (II), (III), (I)
   (d) (II) = (III), (I)
   (e) (I) = (II) = (III)

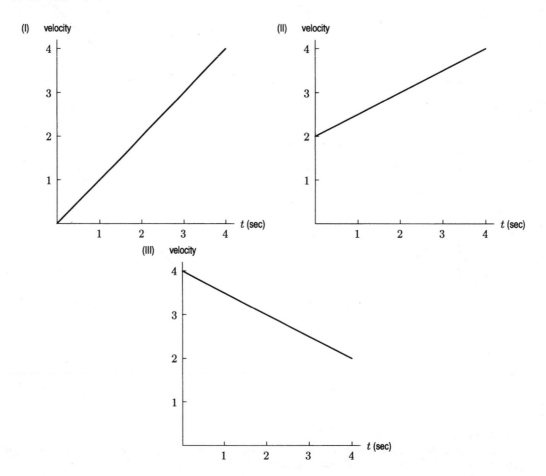

ANSWER:

(d). The distance traveled in this situation is the area under the graph. These areas are (I) $(1/2)(4)(4) = 8$, (II) $(1/2)(4)(4 + 2) = 12$, (III) $(1/2)(4)(4 + 2) = 12$.

COMMENT:

You could mimic this problem using $y = 4 - x$, $y = 4 - x/2$, and $y = 1 + 3x$.

2. Figure 6.5 contains a graph of velocity versus time. Which of the following could be an associated graph of position versus time?

(a) (I)   (b) (II)   (c) (III)
(d) (IV)   (e) (I), (IV)   (f) (II), (III)

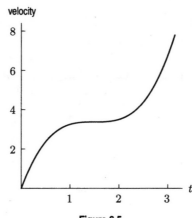

Figure 6.5

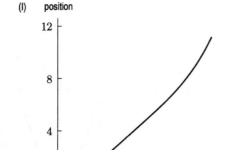

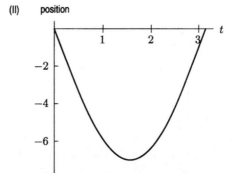

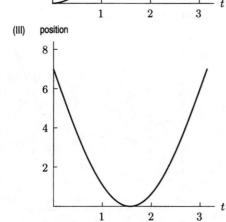

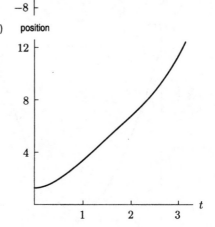

ANSWER:

(e). Because the velocity is positive for the interval shown, the position versus time graph must be increasing for the interval. Notice graphs (I) and (IV) differ by a vertical shift and both are possible.

COMMENT:

You could start with simpler problems using $y = 1$, then $y = x$, and finally $y = x^2$.

3. Figure 6.6 contains a graph of velocity versus time. Which of (a)–(d) could be an associated graph of position versus time?

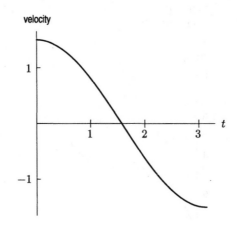

Figure 6.6

(a) position

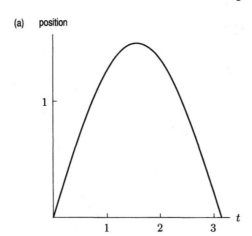

(b) position

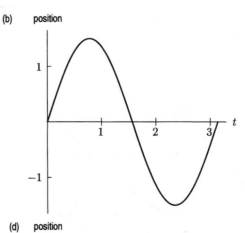

(c) position

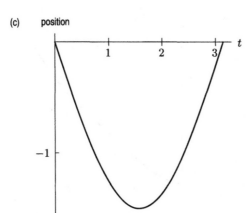

(d) position

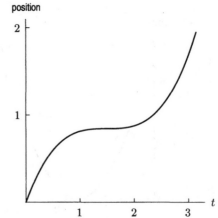

ANSWER:

(a). Because the velocity goes from a positive value to a negative value at $x \approx 1.6$, the position will be a maximum there.

COMMENT:

You could elaborate why each of the other choices has properties which exclude it.

ConcepTests and Answers and Comments for Section 6.4

1. If the graph of f is given in Figure 6.7, then which of (a)–(d) is the graph of $\displaystyle\int_{-3}^{x} f(t)\, dt$ for $-3 < x < 2$?

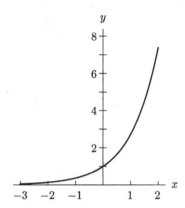

Figure 6.7

(a)
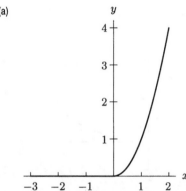

(b)
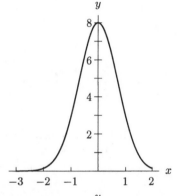

(c)
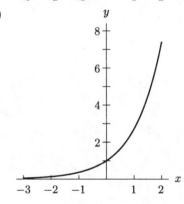

(d)
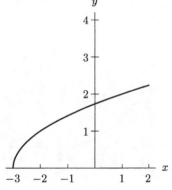

ANSWER:

(c). Because $\displaystyle\int_{-3}^{-3} f(t)\, dt = 0$, the point $(-3, 0)$ is on the graph of $\displaystyle\int_{-3}^{x} f(t)\, dt$. Because the graph of f is positive and increasing. $\displaystyle\int_{-3}^{x} f(t)\, dt$ will be increasing and concave up.

COMMENT:

Each choice could be examined in detail to show why it is not appropriate.

2. If the graph of f is given in Figure 6.8, then which of (a)–(d) is the graph of $\displaystyle\int_2^x f(t)\,dt$ for $0 \le x \le 4$?

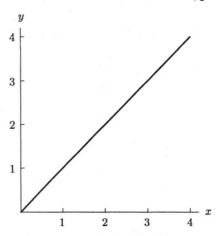

Figure 6.8

(a)
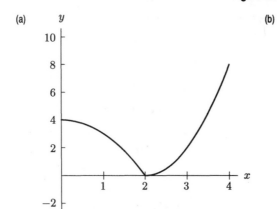

(b)
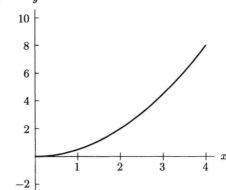

(c)
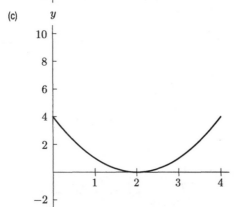

(d)
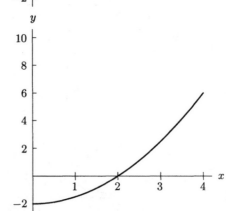

ANSWER:

(d). Because $\displaystyle\int_2^2 f(t)\,dt = 0$, the point $(2,0)$ is on the graph of $\displaystyle\int_2^x f(t)\,dt$. Because $\displaystyle\int_2^0 f(t)\,dt = -\int_0^2 f(t)\,dt$

and $\displaystyle\int_0^2 f(t)\,dt$ is positive, then $\displaystyle\int_2^0 f(t)\,dt$ is negative.

COMMENT:

This example is useful showing the value of interchanging limits.

3. If the graph of f is given in Figure 6.9, then which of (a)–(d) is the graph of $\displaystyle\int_1^x f(t)\,dt$ for $0 \le x \le 3$?

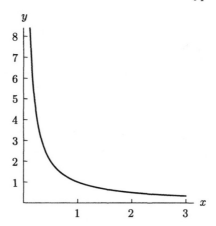

Figure 6.9

(a)
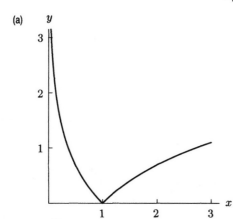

(b)
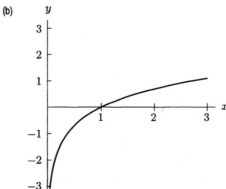

(c)
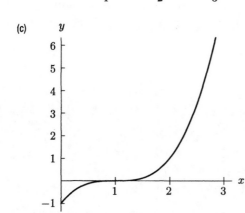

(d)
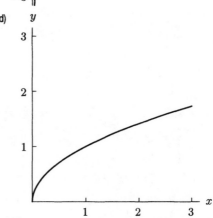

ANSWER:

(b). Because $\displaystyle\int_1^1 f(t)\,dt = 0$, the graph of $\displaystyle\int_1^x f(t)\,dt$ will contain the point $(1,0)$. Because $f(x)$ is positive and $\displaystyle\int_1^0 f(t)\,dt = -\int_0^1 f(t)\,dt$, then $\displaystyle\int_1^0 f(t)\,dt < 0$. f is a decreasing function, so $\displaystyle\int_1^x f(t)\,dt$ will be concave down.

COMMENT:

Have students justify the last statement in the answer.

Chapter Seven

ConcepTests and Answers and Comments for Section 7.1 ————————————

For Problems 1–3, which of the integrals can be converted to one of the form $\int w^n \, dw$ by a substitution, where n is a constant?

1. (a) $\int x \sin(x^2) \, dx$

   (b) $\int \frac{1}{x \ln x} \, dx$

   (c) $\int \frac{1}{\tan x} \, dx$

   (d) $\int x^2 (x^3 + 3)^6 \, dx$

   ANSWER:

   (b), (c), and (d). In (b) $w = \ln x$ yields $\int w^{-1} \, dw$, in (c) $w = \sin x$ yields $\int w^{-1} \, dw$, and in (d) $w = x^3 + 3$ yields $\frac{1}{3} \int w^6 \, dw$.

   COMMENT:

   While $w = x^2$ is a useful substitution for (a), the result is not of the form $\int w^n \, dw$.

2. (a) $\int \frac{4x^3 + 3}{\sqrt{x^4 + 3x}} \, dx$

   (b) $\int \frac{e^x - e^{-x}}{(e^x + e^{-x})^3} \, dx$

   (c) $\int \frac{2x}{x^2 + 1} \, dx$

   (d) $\int \frac{\sin x}{x} \, dx$

   ANSWER:

   (a), (b), and (c). In (a) $w = x^4 + 3x$ yields $\int w^{-1/2} \, dw$, in (b) $w = e^x + e^{-x}$ yields $\int w^{-3} \, dw$, in (c) $w = x^2 + 1$ yields $\int w^{-1} \, dw$.

   COMMENT:

   There is no substitution that converts (d) into a function which has an elementary antiderivative.

3. (a) $\int x^{16} (x^{17} + 16)^{16} \, dx$

   (b) $\int x^{16} (x^{17} + 16x)^{16} \, dx$

   (c) $\int \frac{18x}{1 + 6x^3} \, dx$

   (d) $\int \frac{e^x}{e^x + 6} \, dx$

   ANSWER:

   (a) and (d). In (a) $w = x^{17} + 16$ yields $\frac{1}{17} \int w^{16} \, dw$, in (d) $w = e^x + 6$ yields $\int w^{-1} \, dw$.

   COMMENT:

   You might note why $w = x^{17} + 16x$ does not work for choice (b).

4. Which of the following integrals yields an arcsine or arctangent function upon integration after an appropriate substitution?

   (a) $\int \frac{x}{4 - x^2} \, dx$

   (b) $\int \frac{4}{4 + x^2} \, dx$

   (c) $\int \frac{x}{\sqrt{4 - x^2}} \, dx$

   (d) $\int \frac{x}{\sqrt{4 - x^4}} \, dx$

   ANSWER:

   (b) and (d). In (b) $w = x/2$ yields an arctangent function while in (d) $w = x^2/2$ yields an arcsine function.

   COMMENT:

   You might note which substitution works in (a) and (c).

ConcepTests and Answers and Comments for Section 7.2

1. For which of the following integrals is integration by parts with $u = x$ and $v' =$ (the rest of the integrand) a reasonable choice?

   (a) $\displaystyle\int x(\ln x)^2 \, dx$

   (b) $\displaystyle\int xe^{x^3} \, dx$

   (c) $\displaystyle\int x \sin x \, dx$

   (d) $\displaystyle\int \frac{x}{\ln x} \, dx$

   ANSWER:

   (c). Calculating this integral using integration by parts requires knowing $\int \cos x \, dx$. In contrast, using parts with $u = x$ on (a) requires the knowledge of $\int (\ln x)^2 \, dx$; (b) requires the knowledge of $\int e^{x^3} \, dx$; (d) requires the knowledge of $\int \frac{1}{\ln x} \, dx$.

   COMMENT:

   Compute (c); the value is $-x \cos x + \sin x$.

ConcepTests and Answers and Comments for Section 7.3

1. Which of the following integrals equals $\int x^5 \cos(x^3) \, dx$ with the given substitution?

   (a) $\int w^5 \cos w \, dw, \quad w = x^3$
   (b) $\int w^{5/3} \cos w \, dw, \quad w = x^3$
   (c) $\frac{1}{3} \int w \cos w \, dw, \quad w = x^3$
   (d) $\int w \cos(w^{3/5}) \, dw, \quad w = x^5$

   ANSWER:

   (c). Only the third substitution gives the original integral.

   COMMENT:

   Follow-up Question. In the other three cases, what is wrong?

2. True or false? Using the given substitution, $\displaystyle\int \frac{e^x + 1}{e^x} \, dx$ equals

   (a) $\displaystyle\int \frac{w}{w-1} \, dw, \; w = e^x + 1$

   (b) $\displaystyle\int -(1 + e^w) \, dw, \; w = -x$

   (c) $\displaystyle\int \frac{w+1}{w} \, dw, \; w = e^x$

   (d) $\displaystyle\int w \, dw, \; w = \frac{e^x + 1}{e^x}$

   ANSWER:

   (b). Only the second integral is equal to the original integral.

   COMMENT:

   Follow-up Question. In the other three cases, what is wrong?

ConcepTests and Answers and Comments for Section 7.4

1. Which of the following integrals can be integrated using partial fractions?

   (a) $\displaystyle\int \frac{1}{x^4 - 3x^2 + 2}\, dx$
   (b) $\displaystyle\int \frac{1}{x^4 + 1}\, dx$

   (c) $\displaystyle\int \frac{1}{x^3 - 4x}\, dx$
   (d) $\displaystyle\int \frac{1}{x^4 + 2x^2 + 2}\, dx$

   ANSWER:

   (a) and (c). In (a) the denominator factors as $(x^2 - 2)(x^2 - 1) = (x - \sqrt{2})(x + \sqrt{2})(x - 1)(x + 1)$, so partial fractions will work. In (c) the denominator factors as $x(x - 2)(x + 2)$, so partial fractions will work.

   COMMENT:

   You might note that in (b) the denominator does not factor into real factors and in (d) the denominator equals $(x^2 + 1)^2 + 1$, which has no real factors.

For Problems 2–9, which of the following techniques is useful in evaluating the integral?

(a) Integration by parts
(b) Partial fractions
(c) Long division
(d) Completing the square
(e) A trig substitution
(f) Other substitutions

2. $\displaystyle\int \frac{x^2}{\sqrt{1 - x^2}}\, dx$

   ANSWER:

   (e). $x = \sin \theta$ is appropriate.

   COMMENT:

   It is interesting to note that students may initially make the substitution $w^2 = 1 - x^2$ so the integral becomes $-\int \sqrt{1 - w^2}\, dw$. However, this integral still requires a trigonometric substitution.

3. $\displaystyle\int \frac{1}{\sqrt{6x - x^2 - 8}}\, dx$

   ANSWER:

   (d) and (e). $6x - x^2 - 8 = 1 - (3 - x)^2$, so $x - 3 = \sin \theta$ is appropriate.

   COMMENT:

   You could point out that completing the square is often useful when using tables of integrals.

4. $\displaystyle\int x \sin x\, dx$

   ANSWER:

   (a). $u = x$, $v' = \sin x$ works with integration by parts.

   COMMENT:

   Follow-up Question. Which technique would be useful if the integrand were $x^2 \sin x$?

   Answer. (a). You will need to use integration by parts twice. Initially $u = x^2$, $v' = \sin x$, and the second time $u = x$, $v' = \cos x$.

5. $\displaystyle\int \frac{x}{\sqrt{1 - x^2}}\, dx$

   ANSWER:

   (e) and (f). $w = \sin x$ and $w = 1 - x^2$ work.

   COMMENT:

   Follow-up Question. Which technique would be useful if the x in the numerator was replaced by x^2?

   Answer. (a) and either (e) or (f). This integral can be evaluated in two steps. Initially, using integration by parts, let $u = x$, $v' = x/\sqrt{1 - x^2}$. In order to finish finding the antiderivative, you will have the integral that is in the Problem 2.

6. $\displaystyle\int \frac{x^2}{1 - x^2}\, dx$

   ANSWER:

   (c) and (b). $\dfrac{x^2}{1 - x^2} = -1 + \dfrac{1}{1 - x^2} = 1 + \dfrac{1}{(1 - x)(1 + x)}$.

   COMMENT:

   Follow-up Question. Which technique would be useful if the x^2 in the numerator was replaced by x?

   Answer. (f). The substitution $w = 1 - x^2$ would be appropriate.

7. $\int (1 + x^2)^{-3/2} \, dx$

ANSWER:

(e). $x = \tan \theta$ works.

COMMENT:

Follow-up Question. Which technique would be useful if there was a multiplicative factor of x in the integrand?

Answer. (f). The substitution $w = 1 + x^2$ would be appropriate.

8. $\int \dfrac{x}{\sqrt{1 - x^4}} \, dx$

ANSWER:

(e) and (f). $x^2 = \sin \theta$ and $w = x^2$ both work.

COMMENT:

Follow-up Question. Which technique would be useful if the 1 under the square root sign was replaced by 16?

Answer. (f). Rewrite the integrand as $x/(4\sqrt{1 - x^4/16})$, and let $w = x^2/4$.

9. $\int \dfrac{1}{1 - x^2} \, dx$

ANSWER:

(b). Since the denominator factors as $(1 - x)(1 + x)$ then partial fractions is appropriate.

COMMENT:

Follow-up Question. Which technique would be useful if the denominator was $16 - x^2$?

Answer. (b). Since the denominator factors as $(4 - x)(4 + x)$ then partial fractions is an appropriate method.

ConcepTests and Answers and Comments for Section 7.5

1. The graph in Figure 7.1 is that of $y = f(x)$ for $0 < x < 2$.

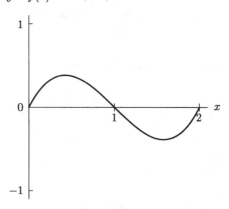

Figure 7.1

For each of the following find the largest interval $a < x < b$ so that $\displaystyle\int_a^b f(x) \, dx$ has the given property.

(a) Both right and trapezoid approximations give overestimates.
(b) Both left and midpoint approximations give overestimates.
(c) Both right and midpoint approximations give underestimates.
(d) Both left and trapezoid approximations give underestimates.

ANSWER:

(a) This occurs where the function is increasing and concave up $a \approx 1.6, b = 2$.
(b) This occurs where the function is decreasing and concave down $a \approx 0.4, b = 1$.
(c) This occurs where the function is decreasing and concave up $a = 1, b \approx 1.5$.
(d) This occurs where the function is increasing and concave down $a = 0, b \approx 0.4$

COMMENT:

You may want to sketch other graphs and ask the same question.

ConcepTests and Answers and Comments for Section 7.6 ━━━━━━

1. (a) For what type of integrand do the Left Sum and Right Sum approximations give the exact value for the integral on any interval?
   (b) For what type of integrand do the Trapezoid and Midpoint rules give the exact value for the integral on any interval?
   ANSWER:

   (a) For any constant function.
   (b) For any linear function.
   COMMENT:
   Follow-up Question. What about Simpson's rule?

2. True or false? If the Trapezoid rule overestimates the value of an integral, then the Midpoint rule underestimates the value of that same integral.
   ANSWER:
   False. Consider $f(x) = |\sin x|$ on $0 \le x \le 2\pi$. Then $\int_0^{2\pi} f(x)\, dx$ is positive, but both the trapezoid and midpoint rule give 0.
   COMMENT:
   Follow-up Question. Under what condition is this statement true?

3. Is it possible to find a function such that all four methods (Left Sum, Right Sum, Midpoint, Trapezoid) overestimate the value of an integral using one subinterval?
   ANSWER:
   Yes. Let $f(x)$ be equal to $-|\sin x|$ on $[0, 2\pi]$. On this interval, using one subinterval, all four methods give a value of 0. The value of the integral is negative, so all four are overestimates.
   COMMENT:
   All four methods also give the same estimate for this example.

ConcepTests and Answers and Comments for Section 7.7 ━━━━━━

1. The graph in Figure 7.2 is that of $y = 2^{-x}$. Which of the following represents the areas of the rectangles considered sequentially?

   (a) $1/2, 3/4, 7/8, 15/16, \ldots$ (b) $1/2, 1/4, 1/8, 1/16, \ldots$
   (c) $1/2, 1/3, 1/4, 1/5, \ldots$ (d) $1, 2, 3, 4, 5, \ldots$

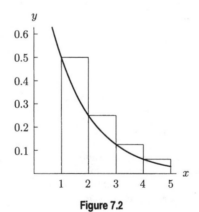

Figure 7.2

ANSWER:
(b). Because the top of all the rectangles lie on $y = 2^{-x}$, their heights will be $1/2, 1/4, 1/8, 1/16, \ldots$. Since the width of each rectangle is 1, this is also the sequence of areas.
COMMENT:
You could point out that this is a left-hand sum for a decreasing function, therefore this sum is greater than the area under the curve.

2. The graph in Figure 7.3 is that of $y = 2^{-x}$ and the area of sequential rectangles is $1/2, 1/4, 1/8, 1/16, \ldots$. What can you conclude?

   (a) The area under the graph of this function from 1 to x as x approaches infinity is finite.
   (b) The area under the graph of this function from 1 to x as x approaches infinity is infinite.
   (c) There is no way to determine whether the area considered is finite or infinite.

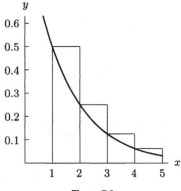

Figure 7.3

ANSWER:

(a). The area of sequential rectangles forms a geometric sequence with a common ratio of $1/2$. Thus the sequence converges. Since the sum of the sequence is an overestimate of the area under the curve, then the area under the curve must also be finite.

COMMENT:

You could also use $s = a_0/(1 - r)$ as the sum to obtain an upper bound.

3. The curve in Figure 7.4 is that of $y = 1/x$ and the widths of the rectangles are $1/32, 1/16, 1/8, 1/4, \ldots$. Which of the following sequences of numbers represents the area of the rectangles in sequential order from left to right?

   (a) $1, 1, 1, 1, \ldots$
   (b) $1/2, 1/4, 1/8, 1/16, \ldots$
   (c) $1/2, 1/2, 1/2, 1/2, \ldots$
   (d) $1/2, 1/3, 1/4, 1/5, \ldots$

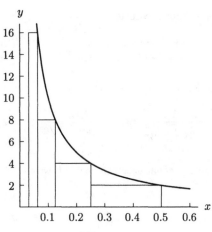

Figure 7.4

ANSWER:

(c). The areas are $16(1/32), 8(1/16), 4(1/8), 2(1/4), \ldots$.

COMMENT:

This process can be continued to obtain the infinite sequence $1/2, 1/2, 1/2, 1/2, \ldots$ for these areas. Because $\displaystyle\int_{1/32}^{\infty} 1/x \, dx$ is larger than the sum of the infinite sequence $1/2, 1/2, 1/2, 1/2, \ldots$, this integral will not be bounded.

4. Which of the following are improper integrals? Explain why.

(a) $\displaystyle\int_1^\infty \frac{\sin x}{x}\, dx$

(b) $\displaystyle\int_4^5 \frac{1}{x}\, dx$

(c) $\displaystyle\int_3^4 \frac{1}{\sin x}\, dx$

(d) $\displaystyle\int_{-3}^3 x^{-1/3}\, dx$

(e) $\displaystyle\int_{-10}^{10} f(x)\, dx$, where $f(x) = \begin{cases} \frac{1}{x} + 2 &,\quad -10 \le x < -1 \\ \frac{1}{x+2} &,\quad -1 \le x \le 10 \end{cases}$

ANSWER:

(a), (c), and (d). For (a) any integral with infinity as a limit is improper, in (c) $\sin \pi = 0$, so the function is not bounded on this interval, and for (d) $x^{-1/3}$ causes the integrand to be unbounded at $x = 0$.

COMMENT:

Some students may profit by seeing another example to point out the difference between having a discontinuity with a finite jump and one with a one-sided limit that grows without bound.

5. Give an example of a function such that $\displaystyle\lim_{t\to\infty} f(t) = 0$ but $\displaystyle\int_1^\infty f(t)\, dt$ diverges.

ANSWER:

$f(t) = 1/t^b$ where $0 < b \le 1$.

COMMENT:

Have students give several examples here.

6. Give an example of a function such that $\displaystyle\lim_{t\to 0^+} f(t) = \infty$ but $\displaystyle\int_0^1 f(t)\, dt$ is finite.

ANSWER:

$f(t) = 1/t^b, 0 < b < 1$.

COMMENT:

Have students give several examples here.

7. Which of the following techniques is useful in evaluating $\displaystyle\int \frac{1}{\sqrt{x}\sqrt{1+\sqrt{x}}}\, dx$?

(a) Integration by parts   (b) Partial fractions   (c) Long division

(d) Completing the square   (e) A trig substitution   (f) Other substitutions

ANSWER:

(f). $w = 1 + \sqrt{x}$ works, since $x = (w-1)^2, dx = 2(w-1)dw$

and

$$\int \frac{1}{\sqrt{x}\sqrt{1+\sqrt{x}}}\, dx = \int \frac{2(w-1)}{(w-1)\sqrt{w}}\, dw = 2\int \frac{dw}{\sqrt{w}}.$$

COMMENT:

Follow-up Question. If this were a definite integral from a to b, what are the limitations on a and b?

Answer. $0 \le a, b, < \infty$.

ConcepTests and Answers and Comments for Section 7.8

1. Without evaluating the integral, state which of the following improper integrals will converge and which will diverge.

   (a) $\displaystyle\int_1^\infty \frac{1}{x^{10}}\,dx$

   (b) $\displaystyle\int_1^\infty x^{-10}\sin x\,dx$

   (c) $\displaystyle\int_1^\infty x^{-1/3}\,dx$

   (d) $\displaystyle\int_1^\infty x^{-1/3}\arctan x\,dx$

   ANSWER:

   (a) $\displaystyle\int_a^\infty \frac{1}{x^b}\,dx$ converges for $a > 0$ and $b > 1$.

   (b) This integrand is always less than or equal to the integrand in (a), so it will converge because the integral in (a) converges.

   (c) $\displaystyle\int_a^\infty \frac{1}{x^b}\,dx$ diverges for $a > 0$ and $b < 1$.

   (d) The arctangent function is always increasing, and for $1 < x < \infty$ has a minimum value of $\pi/4$. Thus this integrand is greater than $(\pi/4)x^{-1/3}$, and $\displaystyle\int_1^\infty \frac{\pi}{4}x^{-1/3}\,dx$ diverges. Thus this integral diverges.

   COMMENT:

   You could ask students to come up with other examples of improper integrals which either converge or diverge but have monotone decreasing integrands. Then repeat the question with oscillating integrands. Also you could ask if it is possible to have a convergent improper integral with a monotone increasing function. (The answer is yes, consider $\int_1^\infty -1/x^2\,dx$.)

Chapter Eight

ConcepTests and Answers and Comments for Section 8.1 ━━━━━━━━━━━━━

1. Which of the following graphs (a)–(d) represents the area under the line shown in Figure 8.1 as a function of x?

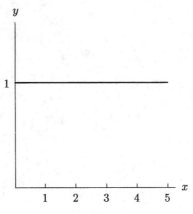

Figure 8.1

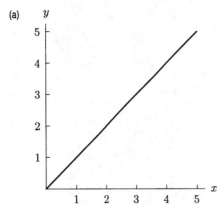

(a)

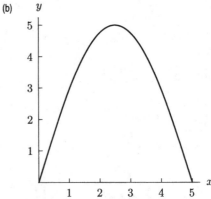

(b)

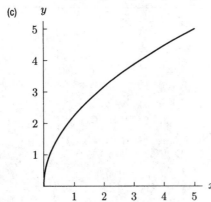

(c)

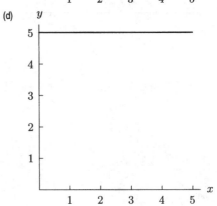

(d)

ANSWER:

(a). Because the graph shown is that of a positive constant, the area will be directly proportional to the length of the interval.

COMMENT:

You could have your students explain why choice (d) is not the area under any given function from 0 to x.

2. Which of the following graphs (a)–(d) represents the area under the line shown in Figure 8.2 as a function of x?

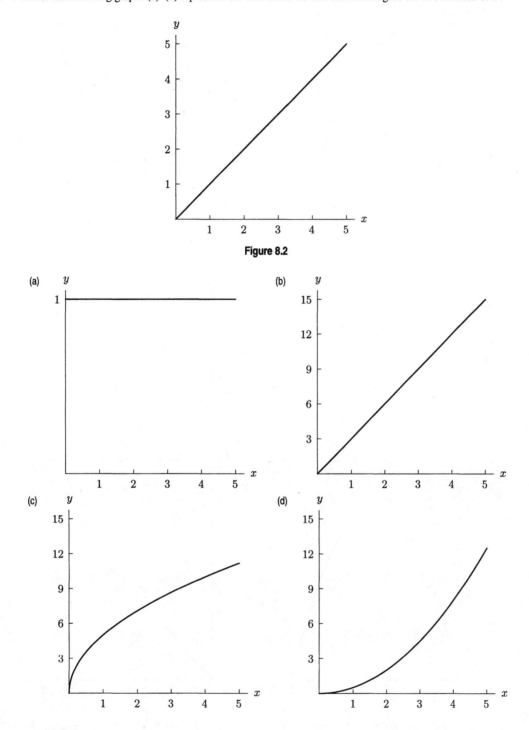

Figure 8.2

ANSWER:

(d). Because the graph in Figure 8.2 is positive and increasing, its antiderivative is increasing and concave up.

COMMENT:

You could ask your students which properties a function must have so that the graph of its area from 0 to x is concave down.

3. Consider the area between the two functions shown in Figure 8.3. Which of the following graphs (a)–(d) represents this area as a function of x?

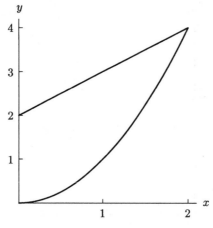

Figure 8.3

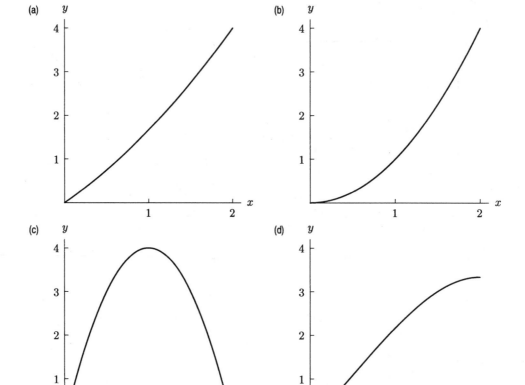

ANSWER:

(d) Because the vertical distance between the two curves continually decreases, the graph of the area between these curves will be increasing and concave down.

COMMENT:

Follow-up Question. How can you change the curves in Figure 8.3 so that the graph representing the area between the curves is horizontal for some interval?

Answer. In order for the graph representing the area between the curves to be horizontal on an interval, the area between the graphs is unchanging. Therefore, on that interval the two curves should be equal.

4. Consider the area between the two functions shown in Figure 8.4. Which of the following graphs (a)–(d) represents this area as a function of x?

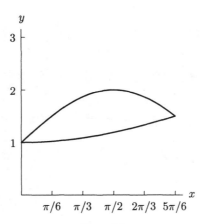

Figure 8.4

(a)

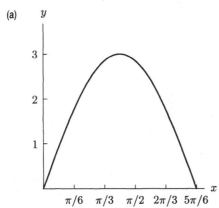

(b)

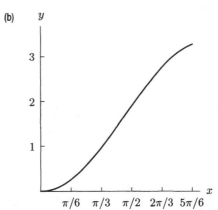

(c)

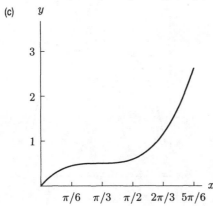

(d)

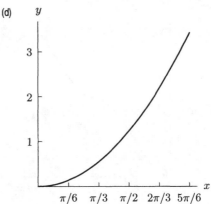

ANSWER:

(b). Because the vertical distance between the two curves increases, and then decreases, the graph of the area between them will first be concave up and then concave down.

COMMENT:

Follow-up Question. What situation would cause the area between the two curves to change from concave down to concave up as in choice (c)?

Answer. When $x = 0$ the curves are far apart and move closer to each other until $x \approx \pi/3$. At this point the curves move farther apart.

5. Consider the area between the two functions shown in Figure 8.5. Which of the following graphs (a)–(d) represents this area as a function of x?

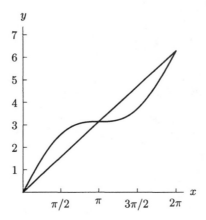

Figure 8.5

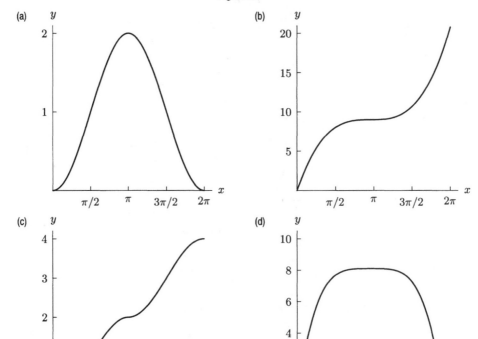

ANSWER:

(c). Because the vertical distance between the two curves increases for $0 < x < \pi/2$ and $\pi < x < 3\pi/2$, the graph of the area will be concave up in these intervals. It will be concave down in the other intervals because there the vertical distance between the given curves decreases.

COMMENT:

Follow-up Question. How is the graph of the integral from 0 to x of the difference between these two functions related to the graph in (c)?

Answer. After $x = \pi$, the graph in (c) needs to be reflected across the line $y = 2$ in order to represent the integral of the difference between the curves from 0 to x.

6. A concrete slab in the form of an isosceles right triangle (the two legs are 5 meters in length) is the base for a tent (see Figure 8.6). The ribs and ceiling of the tent are made of poles such that a cross-section of the tent forms a square. (This cross-section is parallel to one leg as it connects the hypotenuse to the other leg.) There are only two tents that have this configuration, one with the ribs attached to the x-axis, and the other with ribs attached to the line $x = 5$. Which of the following statements is true?

(a) The tent with ribs fastened to the x-axis and hypotenuse will contain the larger volume.
(b) The tent with ribs fastened to the line $x = 5$ and hypotenuse will contain the larger volume.
(c) The tents in choices (a) and (b) will contain the same volume.
(d) We need more information to determine which will contain the larger volume.

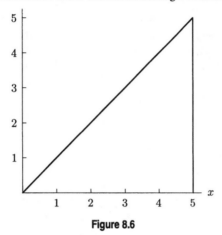

Figure 8.6

ANSWER:
(c) This can be seen using symmetry considerations.
COMMENT:
You may have students who need to actually set up the two integrals before believing the answer.

7. Which of the following gives the volume of the tent in Question 6?

(a) $\displaystyle \int_0^5 x^2 \, dx$

(b) $\displaystyle \int_0^5 (5 - y)^2 \, dy$

(c) Either answer in (a) or (b)

(d) None of the above

ANSWER:
(c). Choice (a) is obtained by slicing along the x-axis, choice (b) by slicing along the line $x = 5$. (Recall that the hypotenuse is given by $y = x$.)
COMMENT:
You can mimic this question and Question 6 by having the cross-section of the tent be an isosceles right triangle or a trapezoid (with two sides perpendicular to the base and the other sides unequal).

8. Now consider a tent over the same slab from Questions 6 and 7 where the ribs are placed parallel to one leg and attached to the hypotenuse so the cross-section forms a semi-circle, with the diameter along the slab. The ratio of the resulting volume to the volume having square cross-sections as in Question 7 is

(a) $\pi/16$

(b) $\pi/8$

(c) $\pi/4$

(d) $\pi/2$

ANSWER:
(b). This can be seen by noting that the volume of a cross-section of width Δx is $(1/2)\pi(\text{radius})^2 \Delta x = (1/2)\pi(x/2)^2 \Delta x$ and the area of the square cross-section (as in Question 7) of width Δx is $x^2 \Delta x$.
COMMENT:
Have the students draw cross-sections to compare the area of a square with a semi-circle with one of the sides as a base to estimate the answer before doing any calculations.

9. A concrete slab in the form of a right triangle with legs of 5 and 10 meters is the base for a tent (see Figure 8.7). The ribs and ceiling of the tent are made of poles such that a cross-section of the tent forms a square. (This cross-section is parallel to one leg as it connects the hypotenuse to the other leg.) There are only two tents that have this configuration, one with the ribs attached to the x-axis, and the other with ribs attached to the line $x = 5$. Which of the following statements is true? Note that the horizontal and vertical scales are different.

   (a) The tent with ribs fastened to the x-axis and hypotenuse will contain the larger volume.
   (b) The tent with ribs fastened to the line $x = 5$ and hypotenuse will contain the larger volume.
   (c) The tents in choices (a) and (b) will contain the same volume.
   (d) We need more information to determine which will contain the larger volume.

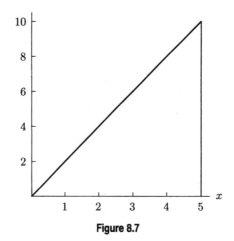

Figure 8.7

   ANSWER:
   (a). The longer span will produce larger square cross-sections so it seems that placing the ribs parallel to the y-axis will produce the larger volume.
   COMMENT:
   You might have students sketch some cross-sections before asking for the answer.

10. The ratio of the volumes of the two cross-sections described in Question 9 is

   (a) 5/4
   (b) 4/3
   (c) 3/2
   (d) 2/1

   ANSWER:
   (d). If the ribs are fastened to the x-axis, we consider slices of width Δx along that axis. The approximate volume of such a slice is $y^2 \Delta x$, which leads to the volume integral

$$\int_0^5 y^2 \, dx = \int_0^5 (2x)^2 \, dx = (4/3)5^3.$$

   If the ribs are fastened to the line $x = 5$, we consider slices of width Δy along that line. The approximate volume of such a slice is $(5 - x)^2 \Delta y$, which leads to the volume integral

$$\int_0^{10} (5 - x)^2 \, dy = \int_0^{10} (5 - y/2)^2 \, dy = (2/3)5^3.$$

   COMMENT:
   You could now ask a similar question for other shaped cross-sections–like semi-circles, triangles, trapezoids, etc.

ConcepTests and Answers and Comments for Section 8.2

1. Imagine taking the enclosed region in Figure 8.8 and rotating it about the x-axis. Which of the following graphs (a)–(d) represents the resulting volume as a function x?

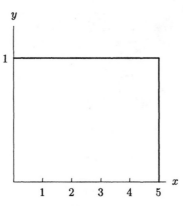

Figure 8.8

(a)
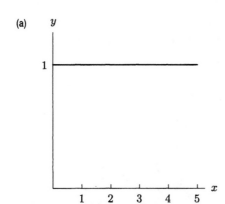

(b)
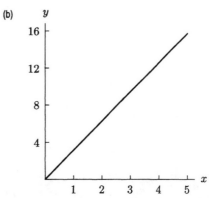

(c)
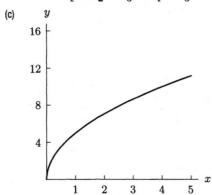

(d)
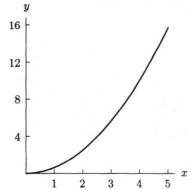

ANSWER:

(b). The volume of a small disc of width Δx is $\pi(1)^2\Delta x$, so the volume integral giving the total volume is $\int_0^x \pi\,dx = \pi x$ where $0 < x < 5$.

COMMENT:

Notice that the volume has the shape of a cylinder, so the volume will be directly proportional to its height.

2. Imagine taking the enclosed region in Figure 8.9 and rotating it about the y-axis. Which of the following graphs (a)–(d) represent the resulting volume as a function of x?

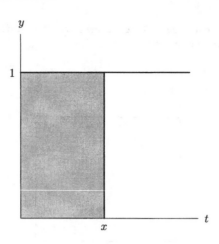

Figure 8.9

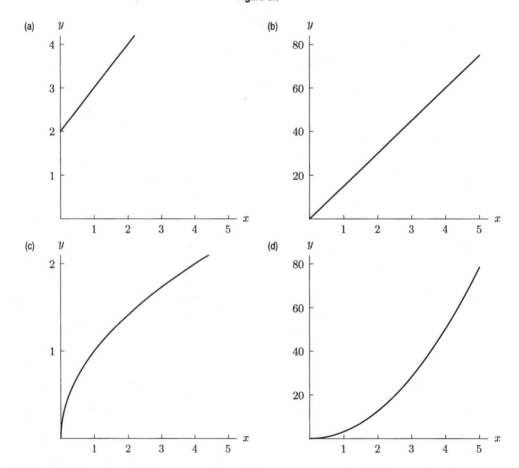

ANSWER:

(d). This gives a cylinder of radius x and height 1, volume equals πx^2.

COMMENT:

You could also point out that slicing along the vertical axis gives a trivial integration.

3. Imagine that the region between the graphs of f and g in Figure 8.11 is rotated about the y-axis to form a solid. Which of the following represents the volume of this solid?

(a) $\displaystyle\int_0^q 2\pi x(f(x) - g(x))\,dx$

(b) $\displaystyle\int_0^q (f(x) - g(x))\,dx$

(c) $\displaystyle\int_0^q \pi(f(x) - g(x))^2\,dx$

(d) $\displaystyle\int_0^q (\pi f^2(x) - \pi g^2(x))\,dx$

(e) $\displaystyle\int_0^q \pi x(f(x) - g(x))\,dx$

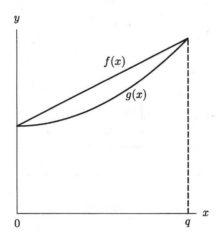

Figure 8.10

ANSWER:

(a). The volume resulting from rotating the area between the two curves bounded by x and $x + \Delta x$ is the difference between the two right cylinders with these two radii. The height of the larger cylinder is $f(x + \Delta x) - g(x + \Delta x)$ and its radius is $x + \Delta x$, while the height of the smaller cylinder is $f(x) - g(x)$, with a radius of x. Thus the difference between these two volumes is $\pi(f(x + \Delta x) - g(x + \Delta x))(x + \Delta x)^2 - \pi(f(x) - g(x))x^2$. If Δx is small, it makes little difference if the functions are evaluated at x or $x + \Delta x$, and if we ignore terms involving $(\Delta x)^2$ in comparison with those involving Δx we obtain $2\pi(f(x) - g(x))x\Delta x$.

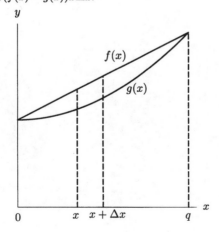

Figure 8.11

COMMENT:

You could also have the students set up the integral giving this volume by slicing along the y-axis. This will involve the use of inverse functions.

4. Imagine rotating the enclosed region in Figure 8.12 about three lines separately: the x-axis, the y-axis, and the vertical line at $x = 6$. This produces three different volumes. Which of the following lists those volumes in order of largest to smallest?

   (a) x-axis; $x = 6$; y-axis
   (b) y-axis; $x = 6$; x-axis
   (c) $x = 6$; y-axis; x-axis
   (d) $x = 6$; x-axis; y-axis
   (e) x-axis; y-axis; $x = 6$
   (f) y-axis; x-axis; $x = 6$

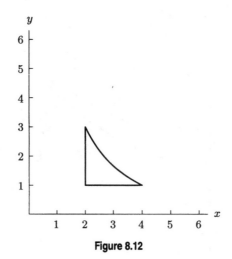

Figure 8.12

ANSWER:

(c). The center of gravity of the enclosed region is at about $(2.5, 1.5)$ and the greater the distance of the line of rotation from the center of gravity, the greater the volume.

COMMENT:

If you want students to calculate three volumes, change the curve in Figure 8.12 to a line giving an isosceles right triangle where $2 < x < 4$. Here the answers are $40\pi/3, 32\pi/3$, and $20\pi/3$.

5. Which of the following graphs (a)–(d) represents the arc length (from the origin) of the curve in Figure 8.13 as a function of x?

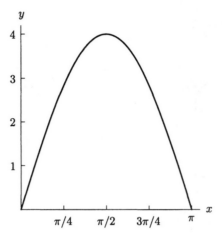

Figure 8.13

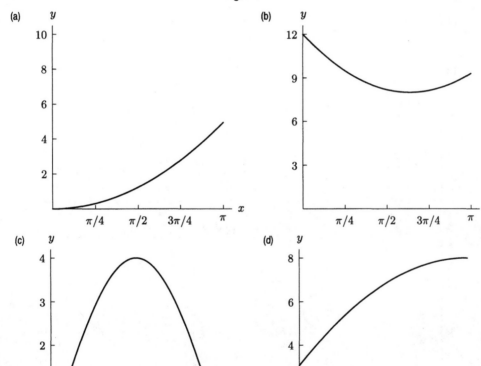

ANSWER:

(a). Arc length is always increasing, which eliminates (c) and (b). The rate of change of arc length is never zero, so there will be no horizontal tangents, this eliminates (d).

COMMENT:

For an easier first problem, consider what happens if $y = \frac{1}{2}x$.

6. Which of the graphs (a)–(d) represents the arc length (from the origin) of the curve in Figure 8.14 as a function of x?

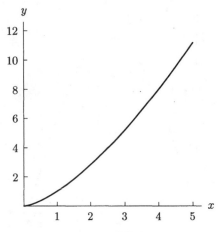

Figure 8.14

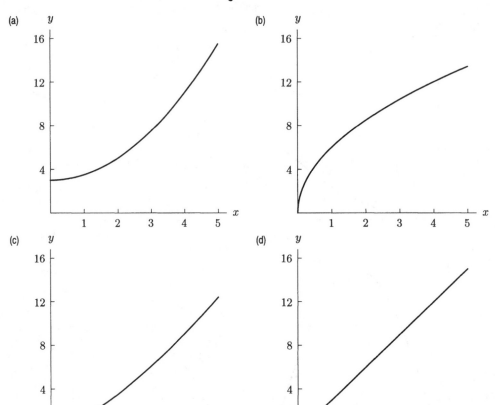

ANSWER:

(c). The arc length curve will start at the origin with non-zero slope. Arc length for an increasing function will have the same concavity as that of the function. This eliminates (b) and (d).

COMMENT:

Have students decide how to prove this. (They should differentiate $\int_{a}^{x} \sqrt{1 + (f'(t))^2}\, dt$ twice.)

ConcepTests and Answers and Comments for Section 8.3

1. Which describes the graph of the equation $r = \theta$?

   (a) Line
   (b) Circle
   (c) Spiral
   (d) Rose

   ANSWER:

   (c). As a point moves along the graph with increasing angle θ, circling the origin counterclockwise, it simultaneously moves away from the origin, creating a spiral. This happens because r increases along with θ.

2. Which describes the graph of the equation $r \sin \theta = 10$?

   (a) Line
   (b) Circle
   (c) Spiral
   (d) Rose

   ANSWER:

   (a). Since $r \sin \theta = y$, the curve is given in Cartesian coordinates by the equation $y = 10$, which is a horizontal line.

3. Which describes the graph of the equation $r = \cos \theta$?

   (a) Line
   (b) Circle
   (c) Spiral
   (d) Rose

   ANSWER:

   (b) and (d).

   Multiplying the equation by r we have $r^2 = r \cos \theta$. Since $r^2 = x^2 + y^2$ and $r \cos \theta = x$, the curve is given in Cartesian coordinates by the equation $x^2 + y^2 = x$. Completing of the square gives $(x - 1/2)^2 + y^2 = (1/2)^2$, the equation of a circle of radius $1/2$ centered at the point $(1/2, 0)$. This is answer (b).

   A rose has equation $r = a \cos n\theta$, which is $r = \cos \theta$ with $a = 1$ and $n = 1$. This is answer (d).

4. Which of the following equations would produce the graph shown in Figure 8.15?

   (a) $r = 8\theta$ (b) $r = 8 \cos \theta$ (c) $r = 8/\cos \theta$ (d) $r = 8 \sin \theta$

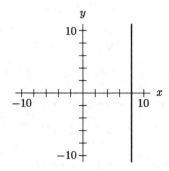

Figure 8.15

ANSWER:

(c). This is the graph of $x = 8$. Since $x = r \cos \theta$, we have $x = r \cos \theta = 8$. Thus, $r = 8/\cos \theta$.

5. Which of the following equations would produce the graph shown in Figure 8.16?

   (a) $r = 2\cos\theta$ (b) $r = 4\cos\theta$ (c) $r = 2\sin\theta$ (d) $r = 4\sin\theta$ (e) $r = 2$

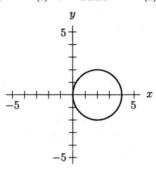

Figure 8.16

ANSWER:

(b). This is circle with a radius of 2 whose center is at the point $(2,0)$. In Cartesian coordinates the equation of the circle is $(x-2)^2 + y^2 = 4$. By expanding the left side of the equation we get $x^2 - 4x + 4 + y^2 = 4$. Simplifying the equation, we get $x^2 + y^2 = 4x$.

Since $x^2 + y^2 = r^2$, and $x = r\cos\theta$, we have $r^2 = 4r\cos\theta$. Since $r \neq 0$, we can divide both sides of the equation by r, giving $r = 4\cos\theta$.

6. Which of the following equations would produce a rose with 8 petals and length 4?

   (a) $r = 4\cos(4\theta)$ (b) $r = 4\cos(8\theta)$ (c) $r = 8\cos(4\theta)$ (d) $r = 8\cos(8\theta)$

   ANSWER:

   (a). The equation $r = a\cos(n\theta)$ produces a rose with length a and n petals if n is odd and $2n$ petals if n is even. To produce a rose with 8 petals and length 4, we let $a = 4$ and $n = 4$.

7. Which of the following represents the area inside the rose $r = 5\sin 2\theta$?

   (a) $A = \dfrac{1}{2}\displaystyle\int_0^{2\pi} 5\sin^2 2\theta\, d\theta$ (b) $A = 2\displaystyle\int_0^{\pi/2} 5\sin^2 2\theta\, d\theta$ (c) $A = \dfrac{1}{2}\displaystyle\int_0^{\pi} (5\sin 2\theta)^2\, d\theta$

   (d) $A = 2\displaystyle\int_0^{\pi/2} (5\sin 2\theta)^2\, d\theta$ (e) $A = 4\displaystyle\int_0^{\pi/2} (5\sin 2\theta)^2\, d\theta$

   ANSWER:

   (d). The equation $r = 5\sin 2\theta$ produces a rose with length 5 and 4 petals.

   The formula for the area inside this curve is: $A = \dfrac{1}{2}\displaystyle\int_0^{2\pi} (5\sin 2\theta)^2\, d\theta$.

   Since this is not one of the choices, we note that in the interval $0 \le \theta \le \pi/2$, one petal is completed. Using this interval, we can find the area of one petal and multiply by 4 to find the area inside the entire curve.

   $$A = 4 \cdot \frac{1}{2}\int_0^{\pi/2} (5\sin 2\theta)^2\, d\theta = 2\int_0^{\pi/2} (5\sin 2\theta)^2\, d\theta.$$

8. Which of the following represents the area inside the rose $r = 5\cos 2\theta$?

   (a) $A = \dfrac{1}{2}\displaystyle\int_0^{2\pi} 5\cos^2 2\theta\, d\theta$ (b) $A = 2\displaystyle\int_0^{\pi/2} 5\cos^2 2\theta\, d\theta$ (c) $A = \dfrac{1}{2}\displaystyle\int_0^{\pi} (5\cos 2\theta)^2\, d\theta$

   (d) $A = 2\displaystyle\int_0^{\pi/4} (5\cos 2\theta)^2\, d\theta$ (e) $A = 4\displaystyle\int_0^{\pi/4} (5\cos 2\theta)^2\, d\theta$

   ANSWER:

   (e). The equation $r = 5\cos 2\theta$ produces a rose with length 5 and 4 petals.

   The formula for the area inside this curve is: $A = \dfrac{1}{2}\displaystyle\int_0^{2\pi} (5\cos 2\theta)^2\, d\theta$.

   Since this is not one of the choices, we note that in the interval $0 \le \theta \le \pi/4$, half of a petal is completed. Using this interval, we can find the area of half a petal and multiply by 8 to find the area inside the entire curve.

   $$A = 8 \cdot \frac{1}{2}\int_0^{\pi/4} (5\cos 2\theta)^2\, d\theta = 4\int_0^{\pi/4} (5\cos 2\theta)^2\, d\theta.$$

9. Which of the following represents the area inside the limaçon $r = 2 + 2\cos\theta$?

(a) $A = \dfrac{1}{2}\displaystyle\int_0^{2\pi} 2 + 2\cos^2\theta\, d\theta$ (b) $A = \displaystyle\int_0^{\pi} 2 + 2\cos^2\theta\, d\theta$ (c) $A = \dfrac{1}{2}\displaystyle\int_0^{\pi} (2 + 2\cos\theta)^2\, d\theta$

(d) $A = \displaystyle\int_0^{\pi} (2 + 2\cos\theta)^2\, d\theta$ (e) None of the above

ANSWER:

(d). The equation $r = 2 + 2\cos\theta$ produces a limaçon with no inner loop. The limaçon is symmetric about the x-axis. The formula for the area inside this curve is: $A = \dfrac{1}{2}\displaystyle\int_0^{2\pi} (2 + 2\cos\theta)^2\, d\theta$.

Since this is not one of the choices, we note that in the interval $0 \le \theta \le \pi$, half of the limaçon is completed. Using this interval, we can find the area of half the limaçon and multiply by 2 to find the area inside the entire curve.

$$A = 2 \cdot \frac{1}{2}\int_0^{\pi} (2 + 2\cos\theta)^2\, d\theta = \int_0^{\pi} (2 + 2\cos\theta)^2\, d\theta.$$

COMMENT:

This limaçon is an example of a cardioid. Have the students explain why this name is appropriate, and then ask under what conditions a limaçon, $(r = a + b\sin\theta, r = a + b\cos\theta)$ is a cardioid. (Answer: $|a| = |b|$.)

10. Which of the following expressions represents the arc length of the curve $r = 4\cos\theta$?

(a) $\displaystyle\int_0^{\pi/2} \sqrt{(-4\sin\theta)^2 + (4\cos\theta)^2}\, d\theta$ (b) $2\displaystyle\int_0^{\pi/2} \sqrt{(-4\sin\theta)^2 + (4\cos\theta)^2}\, d\theta$

(c) $2\displaystyle\int_0^{\pi} \sqrt{(-4\sin\theta)^2 + (4\cos\theta)^2}\, d\theta$ (d) $\displaystyle\int_0^{2\pi} \sqrt{(-4\sin\theta)^2 + (4\cos\theta)^2}\, d\theta$

(e) $2\displaystyle\int_0^{2\pi} \sqrt{(-4\sin\theta)^2 + (4\cos\theta)^2}\, d\theta$

ANSWER:

(b). For a curve of the form $r = f(\theta)$

$$\text{Arc length} = \int_\alpha^\beta \sqrt{(f'(\theta))^2 + (f(\theta))^2}\, d\theta.$$

Since this curve, which is a circle, is completed in the interval $0 \le \theta \le \pi$, the limits of integration are 0 and π.

We can find the length of the arc by using the formula

$$\text{Arc length} = \int_0^{\pi} \sqrt{(-4\sin\theta)^2 + (4\cos\theta)^2}\, d\theta.$$

Alternatively, we can find the arc length using the limits 0 and $\pi/2$, and double this answer.

$$\text{Arc length} = 2\int_0^{\pi/2} \sqrt{(-4\sin\theta)^2 + (4\cos\theta)^2}\, d\theta.$$

Note that the arc length we are seeking is the arc length of a circle whose radius is 2. Thus, we know the answer must be 4π.

11. Which of the following equations represents the area inside the graph of $r = 4$ and outside the graph of $r = 4 - 4\sin\theta$?

(a) $A = 16\pi - \dfrac{1}{2}\displaystyle\int_0^{2\pi} (4 - 4\sin\theta)^2\, d\theta$ (b) $A = 8\pi - \displaystyle\int_0^{2\pi} (4 - 4\sin\theta)^2\, d\theta$

(c) $A = 16\pi - \dfrac{1}{2}\displaystyle\int_0^{\pi} (4 - 4\sin\theta)^2\, d\theta$ (d) $A = 8\pi - \displaystyle\int_0^{\pi} (4 - 4\sin\theta)^2\, d\theta$

(e) $A = 8\pi - \displaystyle\int_0^{\pi/2} (4 - 4\sin\theta)^2\, d\theta$

ANSWER:

(e). The equation $r = 4$ produces a circle with a center at the origin and radius of 4. The equation $r = 4 - 4\sin\theta$ produces a limaçon that is symmetric about the y-axis. See Figure 8.17.

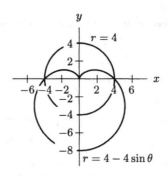

Figure 8.17

We need to find the area of the portion of the circle that is not contained in the graph of the limaçon. Thus, we want to take the area of half the circle and subtract the area enclosed by the top part of the limaçon. (Note that the other half of the circle lies entirely inside the limaçon.)

The area of the circle is 16π. Thus, the area of the semicircle is 8π.

As the limaçon is traced out in the first and second quadrants, the area formed can be determined by the formula:

$$A = \frac{1}{2} \int_0^\pi (4 - 4\sin\theta)^2 \, d\theta.$$

The area that we need can be determined from the formula:

$$A = 8\pi - \frac{1}{2} \int_0^\pi (4 - 4\sin\theta)^2 \, d\theta.$$

This is not an answer choice.

We notice that half of the area that we need is formed in the first quadrant. Thus, if we double this area, we will obtain our answer.

$$A = 2 \cdot \left(4\pi - \frac{1}{2} \int_0^{\pi/2} (4 - 4\sin\theta)^2 \, d\theta \right).$$

We can simplify this and obtain:

$$A = 8\pi - \int_0^{\pi/2} (4 - 4\sin\theta)^2 \, d\theta.$$

COMMENT:

This limaçon is an example of a cardioid, a limaçon ($r = a + b\sin\theta, r = a + b\cos\theta$) for which $|a| = |b|$.

12. Which of the following equations represents the area inside the graph of $r = 3\sin\theta$ and outside the graph of $r = 1 + \sin\theta$?

(a) $A = \dfrac{1}{2} \displaystyle\int_{\pi/6}^{\pi/2} (3\sin\theta)^2 - (1 + \sin\theta)^2 \, d\theta$ (b) $A = \displaystyle\int_{\pi/6}^{\pi/2} (3\sin\theta)^2 - (1 + \sin\theta)^2 \, d\theta$

(c) $A = \dfrac{1}{2} \displaystyle\int_0^\pi (3\sin\theta)^2 - (1 + \sin\theta)^2 \, d\theta$ (d) $A = \displaystyle\int_0^\pi (3\sin\theta)^2 - (1 + \sin\theta)^2 \, d\theta$

(e) $A = \dfrac{1}{2} \displaystyle\int_0^{2\pi} (3\sin\theta)^2 - (1 + \sin\theta)^2 \, d\theta$

ANSWER:

(b). The circle $r = 3\sin\theta$ is completed in the interval $0 \leq \theta \leq \pi$. Since both graphs are symmetric about the y-axis, the region between the graphs is also symmetric about the y-axis. See Figure 8.18.

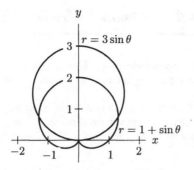

Figure 8.18

Thus we can find the area in the first quadrant and double it. We have to find the point where the two curves intersect in the first quadrant.

By solving the equation $3 \sin \theta = 1 + \sin \theta$, we obtain $\theta = \pi/6$. Thus, the limits of integration are $\pi/6$ and $\pi/2$. The area can be obtained by using the formula:

$$A = 2 \cdot \frac{1}{2} \int_{\pi/6}^{\pi/2} (3 \sin \theta)^2 - (1 + \sin \theta)^2 \, d\theta.$$

By simplifying, we obtain:

$$A = \int_{\pi/6}^{\pi/2} (3 \sin \theta)^2 - (1 + \sin \theta)^2 \, d\theta.$$

ConcepTests and Answers and Comments for Section 8.4

1. A rod of length 3 meters and density $\delta(x) = 2 + \cos x$ grams/meter is positioned along the x-axis with its left end at the origin.

   (a) Where is the rod most dense?
   (b) Where is the rod least dense?
   (c) Is the center of mass of this rod closer to the origin, or closer to $x = 3$?
   (d) What is the total mass of the rod?
   (e) Where is the center of mass of the rod?

   ANSWER:

   (a) The function describing the density is decreasing for $0 \leq x \leq 3$. (Note its derivative, $-\sin x$ is negative for this interval.) So it is most dense at $x = 0$.
   (b) It is least dense at $x = 3$.
   (c) The center of mass is closer to the origin.
   (d) The total mass of the rod is given by

   $$\int_0^3 (2 + \cos x) \, dx = (2x + \sin x)\big|_0^3 = 6 + \sin 3 \approx 6.14 \text{ gm}.$$

   (e) The center of mass is given by

   $$\frac{1}{\text{mass}} \int_0^3 x(2 + \cos x) \, dx = \frac{1}{6.14}(x^2 + x \sin x + \cos x)\big|_0^3 = \frac{1}{6.14}(9 + 3\sin 3 + \cos 3 - 1) \approx 1.21 \text{ m}.$$

   COMMENT:
   You could give a different density function and ask the same set of questions, or simply increase the length of the rod to 6 meters (then you could use symmetry arguments as well).

ConcepTests and Answers and Comments for Section 8.5 ▬▬▬▬▬▬

1. A fuel tank 10 ft tall has a horizontal trapezoidal cross-section with 5 ft sides at two opposite ends and 6 ft and 12 ft at the other ends. Calculate the work done in pumping out the fuel (of density ρ lb/ft^3) if the tank is half full.

    ANSWER:
    Consider a cross-sectional slice of thickness Δh of the tank, h feet from the bottom. The area of this cross-section is $(4)(6+12)/2 = 36$ ft^2, with a corresponding volume of the slice of $36\Delta h$ ft^3. This volume must be lifted a total of $10 - h$ feet, and the corresponding work done on this slice is $36\rho(10 - h)\Delta h$. Adding up all the similar cross-sections from $h = 0$ to $h = 5$ gives

$$\text{Work done} = \int_0^5 36\rho(10 - h)\,dh = 36\rho\left(10h - h^2/2\right)\Big|_0^5 = 1350\rho \text{ foot-pounds.}$$

    COMMENT:
    You could point out that for any cross-section with area, ΔA, the work done for such a problem is $(\rho)(\Delta A)(10-h)$.

ConcepTests and Answers and Comments for Section 8.7 ▬▬▬▬▬▬

1. A population density function is given by $p(x) = \begin{cases} ae^x & , \quad -\infty < x \le 1 \\ 0 & , \quad x > 1. \end{cases}$

    (a) Compute the value of a.
    (b) Find the cumulative distribution function $P(x)$.
    (c) For what value of x is the fraction of the population less than $\frac{1}{2}$?

    ANSWER:

    (a) Since $\int_{-\infty}^{\infty} p(x)\,dx = 1$, we need $\int_{-\infty}^1 ae^x\,dx = 1$. This gives $ae^x\,|_{-\infty}^1 = 1$, so $a = 1/e$.

    (b) $P(x) = \int_{-\infty}^x p(x)\,dx = \begin{cases} e^{x-1} & , \quad x \le 1 \\ 1 & , \quad x > 1. \end{cases}$

    (c) $P(x) = 1/2$ will occur when $e^{x-1} = 1/2$, giving $x = 1 - \ln 2$.

    COMMENT:
    You might want to repeat this question with a density function given by $p(x) = ae^{-x^2}$, $-\infty < x < 0$, and 0 for $x \ge 0$. One purpose of this question is to have them solve part (a) and (c) by using a table or trial and error with the limits of integration.

2. A population density function is given by $p(x) = \begin{cases} x^2/2 & , \quad -1 \le x < 0 \\ a(1 - e^{-x}) & , \quad 0 \le x \le 1 \\ 0 & , \quad \text{otherwise.} \end{cases}$

    (a) Compute the value of a.
    (b) Find the cumulative distribution function $P(x)$.
    (c) For what value of x is the percentage of the population equal to $1/12$?

    ANSWER:

    (a) Since $\int_{-\infty}^{\infty} p(x)\,dx = 1$, we need $\int_{-1}^0 (x^2/2)\,dx + \int_0^1 a(1 - e^{-x})\,dx = 1$. This gives

$$(1/6)x^3\Big|_{-1}^0 + a(x + e^{-x})\Big|_0^1 = 1,$$

    so $a = 5e/6$.

    (b) $P(x) = \int_{-\infty}^x p(x)\,dx = \begin{cases} 0 & , \quad x < -1 \\ \frac{1}{6}(x^3 + 1) & , \quad -1 \le x < 0 \\ \frac{1}{6} + \frac{5e}{6}(x + e^{-x} - 1) & , \quad 0 \le x \le 1 \\ 1 & , \quad x > 1. \end{cases}$

(c) $P(x) = \dfrac{1}{12}$ will occur for $x < 0$, so solve $\dfrac{1}{6}(x^3 + 1) = 1/12$, giving $x = -\left(\dfrac{1}{2}\right)^{1/3}$.

COMMENT:

You could repeat part (c) by asking for the fraction of the population equal to 0.95. Here the proper range of the piecewise solution to part (b) must be determined, and then perhaps a look at the graph (using a trace function) to determine the answer.

ConcepTests and Answers and Comments for Section 8.8

1. A density function is given by $p(x) = \begin{cases} x^2/2 & , \quad -1 \leq x < 0 \\ a(1 - e^{-x}) & , \quad 0 \leq x \leq 1 \\ 0 & , \quad \text{otherwise.} \end{cases}$

   (a) Compute the value of a.
   (b) Find the cumulative distribution function $P(x)$.
   (c) From the graph of $p(x)$, do you expect the mean to be positive or negative?
   (d) Find the exact value of the mean.

   ANSWER:

   (a) Since $\displaystyle\int_{-\infty}^{\infty} p(x)\,dx = 1$, we need $\displaystyle\int_{-1}^{0} \frac{x^2}{2}\,dx + \int_{0}^{1} a(1 - e^{-x})\,dx = 1$. This gives

   $$(1/6)x^3\big|_{-1}^{0} + a(x + e^{-x})\big|_{0}^{1} = 1, \text{ so } a = 5e/6.$$

   (b) $P(x) = \displaystyle\int_{-\infty}^{x} p(x)\,dx = \begin{cases} 0 & , \quad x < -1 \\ (1/6)(x^3 + 1) & , \quad -1 \leq x < 0 \\ (1/6)x^3 + (5e/6)(x + e^{-x} - 1) & , \quad 0 \leq x \leq 1 \\ 1 & , \quad x > 1 \end{cases}$

   (c) It looks like the mean will be positive.

   (d)

   $$\bar{x} = \int_{-\infty}^{\infty} xp(x)\,dx = \int_{-1}^{0} (x^3/2)\,dx + \int_{0}^{1} (5e/6)(x - xe^{-x})\,dx$$

   $$= -1/8 + \left((5e/6)(\frac{x^2}{2} + xe^{-x} + e^{-x})\right)\bigg|_{0}^{1} = 37/24 - 5e/12 \approx 0.409$$

   COMMENT:

   Parts (a) and (b) are the same as Problem 2 on page 183. You could introduce this section by asking these questions for a standard distribution–say Gaussian or normal.

Chapter Nine

ConcepTests and Answers and Comments for Section 9.1 ─────────────

1. Which is the value of s_6 for the recursively defined sequence $s_n = s_{n-1} + 2 + (-1)^n$ for $n > 1$ and $s_1 = 2$?

   (a) 6   (b) 8   (c) 11   (d) 13

   ANSWER:

   (d). The sequence begins $2, 5, 6, 9, 10, 13, \ldots$. The jump from s_{n-1} to s_n is by $2 + (-1)^n = 3$ if n is even and by $2 + (-1)^n = 1$ if n is odd.

2. Which of the sequences I–IV is monotone and bounded for $n \geq 1$?

   I. $s_n = 10 - \dfrac{1}{n}$ II. $s_n = \dfrac{10n + 1}{n}$ III. $s_n = \cos n$ IV. $s_n = \ln n$

   (a) I
   (b) I and II
   (c) II and IV
   (d) I, II, and III

   ANSWER:

   (b). Sequence (I) is monotone increasing, bounded between 9 and 10.

   Sequence (II) is monotone decreasing and bounded between 10 and 11.

   Sequence (III) is bounded between -1 and 1 but it is not monotone because it begins $0.54, -0.42, -0.99, -0.65, \ldots$, which contains both a jump down, from 0.54 to -0.42 and a jump up, from -0.99 to -0.65. Terms in a monotone sequence always jump in the same direction.

   Sequence (IV) is monotone increasing but is not bounded.

3. Which of the sequences I–IV converges?

   I. $s_n = \dfrac{\cos n}{n}$ II. $s_n = \cos\left(\dfrac{1}{n}\right)$ III. $s_n = ne^{-n}$ IV. $s_n = \dfrac{e^n}{n}$

   (a) I and III
   (b) I, II, and III
   (c) II and IV
   (d) All of them

   ANSWER:

   (b). Sequence (I) converges to zero, sequence (II) converges to 1, sequence (III) converges to zero and sequence (IV) diverges.

4. Match each of the sequences (a)–(d) with one of the statements (I)–(IV).

   (a) 2^n (b) $2^{1/n}$ (c) $(1/2)^n$ (d) $(-1/2)^n$

   I   Converges but not monotonically.
   II Converges, monotonically increasing.
   III Converges, monotonically decreasing.
   IV Diverges.

   ANSWER:

   (a) IV
   (b) III
   (c) III
   (d) I

   COMMENT:

   Ask students to supply an example for the statement which is not matched.

5. In each case, insert $\leq$ or $\geq$ between the terms of the two sequences and use the relationship to decide the convergence of both sequences.

(a) 1.5^n and $(1+\sqrt{n})^n$

(b) 0.5^n and $\left(\dfrac{1}{1+n}\right)^n$

(c) $1-\dfrac{1}{n}$ and $1-\dfrac{1}{1+n}$

ANSWER:

(a) Since $1.5 < 1+\sqrt{n}$ for $n \geq 1$, we have
$$1.5^n < (1+\sqrt{n})^n.$$
Since 1.5^n diverges, so does $(1+\sqrt{n})^n$.

(b) Since
$$0.5 = \frac{1}{2} \geq \frac{1}{1+n} > 0,$$
for $n \geq 1$, we have
$$0.5^n \geq \left(\frac{1}{1+n}\right)^n > 0.$$
Since 0.5^n converges to 0, so does $(1/(1+n))^n$.

(c) Since $1/n > 1/(n+1)$, we have
$$1-\frac{1}{n} < 1-\frac{1}{n+1} < 1.$$
Since $1-1/n$ converges to 1, so does $1-1/(n+1)$.

COMMENT:

Ask students to explain why it is important that the inequalities go the way they do. What conclusions could have been drawn if the inequalities had been the other way round?

ConcepTests and Answers and Comments for Section 9.2

1. Which of the following are geometric series?

(a) $1 - 1/2 + 1/4 - 1/6 + \cdots$

(b) $2y - 6y^3 + 18y^5 - 54y^7 + \cdots$

(c) $1/2 + 2/3 + 3/4 + 5/6 + \cdots$

(d) $x + x^2 + x^4 + x^7 + \cdots$

ANSWER:

(b). This has a common ratio of $-3y^2$, the rest are not geometric series.

COMMENT:

Follow-up Question. For what values of y will this series converge?

Answer. The series converges for $y^2 < 1/3$, and it converges to $2y/(1+3y^2)$.

2. Which of the following geometric series converge?

(I) $20 - 10 + 5 - 2.5 + \cdots$

(II) $1 - 1.1 + 1.21 - 1.331 + \cdots$

(III) $1 + 1.1 + 1.21 + 1.331 + \cdots$

(IV) $1 + y^2 + y^4 + y^6 + \cdots$, for $-1 < y < 1$

(a) (I) only

(b) (IV) only

(c) (I) and (IV)

(d) (II) and (IV)

(e) None of the other choices is correct.

ANSWER:

(c). The common ratio in series (I) and (IV) is between -1 and 1. The common ratio in (II) and (III) is greater than one.

COMMENT:

You might point out that the series (II) and (III) must diverge since $\lim_{n \to \infty} |c_n| \neq 0$, where each series can be represented as $\displaystyle\sum_{n=0}^{\infty} c_n$.

3. List the following infinite geometric series in order of increasing values of their sum.

   (I) $5 + 5/3 + 5/9 + 5/27 + \cdots$
   (II) $3 + 2 + 4/3 + 8/9 + \cdots$
   (III) $10 - 5 + 5/2 - 5/4 + \cdots$
   (IV) $4 + 3 + 9/4 + 27/16 + \cdots$

   (a) (I) (II) (III) (IV)
   (b) (II) (I) (IV) (III)
   (c) (III) (I) (II) (IV)
   (d) (IV) (II) (I) (III)

   ANSWER:

   (c). The sums of these series are (I) $= 7.5$, (II) $= 9$, (III) $= 20/3 = 6.667$, and (IV) $= 16$.

   COMMENT:

   As a warm-up question, you could ask the students to identify the ratio for each infinite series.

4. A football team starts its offense 10 yards from the goal line. Each play they gain half of the remaining distance to the goal. Assuming that they continue this process indefinitely, will they ever cross the goal line?

   ANSWER:

   Technically, they will never cross the goal line. In the first play they gain 5 yards, in the second play they gain 2.5 yards, then 1.25 yards and so on. Each time they stop short of the goal line.

   COMMENT:

   Follow-up Question. Find a formula for the total yardage the ball has moved after n plays.
   Answer. The ball has moved $10(1 - (1/2)^n)$ yards after n plays.

5. Which of the following are geometric series? For those that are geometric series, determine whether or not they converge.

   (a) $\dfrac{1}{2} + \dfrac{1}{4} + \dfrac{1}{8} + \dfrac{1}{16} + \cdots$

   (b) $\displaystyle\sum_{n=0}^{\infty} \dfrac{1}{e^2}$

   (c) $1 + \dfrac{1}{2} + \dfrac{1}{3} + \dfrac{1}{4} + \cdots$

   (d) $2 - 2y^2 + 2y^4 - 2y^6 + \cdots$

   (e) $\displaystyle\sum_{n=1}^{\infty} \dfrac{1}{n^2}$

   ANSWER:

   (a) This series is geometric and converges since the common ratio equals $1/2$.
   (b) This series is geometric but diverges since the common ratio equals 1.
   (c) This series is not geometric; it is the harmonic series.
   (d) This series is geometric with a common ratio of $-y^2$. It therefore converges if and only if $|y| < 1$.
   (e) This series is not geometric.

   COMMENT:

   Follow-up Question. For each geometric series that converges, find the sum.
   Answer. Series (a) converges to 1. Series (d) converges to $2/(1 + y^2)$.

6. True or False: $\displaystyle\sum_{n=1}^{\infty} \dfrac{1}{n^2} = \dfrac{1}{1 - (1/n^2)}$.

   ANSWER:

   False, this is not a geometric series; therefore, we can not use the formula for an infinite geometric series to find the sum of the series.

   COMMENT:

   You may want to point out to students that if a series in n converges, it must converge to a number independent of n.

ConcepTests and Answers and Comments for Section 9.3

1. You stand one meter from a wall and walk toward the wall, covering one-half the remaining distance to the wall with each step. Which of the following sequences of numbers represents the remaining distances in meters?

   (a) $1/2, 3/4, 7/8, 15/16, \ldots$
   (b) $1/2, 1/4, 1/8, 1/16, \ldots$
   (c) $1/2, 1/3, 1/4, 1/5, \ldots$
   (d) $1, 2, 3, 4, \ldots$

   ANSWER:

   (b). Your first step will take you $1/2$ meter, leaving $1/2$ meter to go. Your second step will be one-half of the remaining distance, or $1/2(1/2) = 1/4$ meter, leaving 1/4 meter to go. Continuing in this manner gives you sequence (b).

   COMMENT:

   What is the sequence of remaining distances if each step takes you $2/3$ of the way toward the wall?

   You could point out that the total distance traveled cannot exceed 1, and the individual distances (per step) will approach 0.

2. Which of the following defines a convergent sequence of partial sums?

   (a) Each term in the sequence is closer to the last term than any two prior consecutive terms.
   (b) Assume that the sequence of partial sums converges to a number, L. Regardless of how small a number you give me, say ϵ, I can find a value of N such that the Nth term of the sequence is within ϵ of L.
   (c) Assume that the sequence of partial sums converges to a number, L. I can find a value of N such that all the terms in the sequence, past the Nth term, are less than L.
   (d) Assume that the sequence of partial sums converges to a number, L. Regardless of how small a number you give me, say ϵ, I can find a value of N such that all the terms in the sequence, past the Nth term, are within ϵ of L.

   ANSWER:

   (d)

   COMMENT:

   You may wish to write this in terms of inequalities.

3. An incorrect definition of a convergent series is *"An infinite series is convergent when the series is bounded, i.e. the sequence of partial sums never goes above or below two specified values"*. Which of the following is a series that fits this definition but is not convergent?

   (a) $\sum \left(\frac{1}{2}\right)^n$
   (b) $\sum \frac{1}{n}$
   (c) $\sum (-1)^n$
   (d) $\sum 1^n$

   ANSWER:

   (c). The partial sums of $\sum_{n=1}^{\infty} (-1)^n$ alternate between -1 and 0, so are bounded above and below, but $\sum_{n=1}^{\infty} (-1)^n$ does not converge.

   COMMENT:

   It may be worth commenting on the convergent properties of the other three series.

4. A sequence of partial sums that is bounded and always increasing is a convergent sequence.

   (a) True
   (b) False

   ANSWER:

   (a)

   COMMENT:

   Follow-up True/False Question. A sequence of partial sums that is bounded and always decreasing is a convergent sequence.

   Answer. True.

5. True of False?

(a) If $\lim\limits_{n \to \infty} a_n = 0$ then $\sum\limits_{n=0}^{\infty} a_n$ converges.

(b) If $\lim\limits_{n \to \infty} a_n \neq 0$ then $\sum\limits_{n=0}^{\infty} a_n$ diverges.

(c) If $\sum\limits_{n=0}^{\infty} a_n$ converges then $\lim\limits_{n \to \infty} a_n = 0$.

(d) If the terms of a series alternate in sign, then the series converges.

ANSWER:

(a) False. For example, $\sum\limits_{n=1}^{\infty} \dfrac{1}{n}$ diverges even though $\lim\limits_{n \to \infty} \dfrac{1}{n} = 0$.

(b) True.

(c) True. We know that if $\sum a_n$ converges, then $\lim\limits_{n \to \infty} a_n = 0$.

(d) False. For example, the terms of the series $\sum\limits_{n=1}^{\infty} (-1)^n n^2$ alternate in sign, but the series diverges.

COMMENT:

Have students find other counterexamples for the statements that are false.

ConcepTests and Answers and Comments for Section 9.4

1. Which test will help you determine if the series converges or diverges?

$$\sum_{k=1}^{\infty} \frac{5+k}{k!}$$

(a) Integral test
(b) Comparison test
(c) Ratio test

ANSWER:

(c). The ratio test is useful if factorials are involved.

COMMENT:

Ask the students if the series converges or diverges.

2. Which test will help you determine if the series converges or diverges?

$$\sum_{k=1}^{\infty} e^{-k}$$

(a) Integral test
(b) Comparison test
(c) Ratio test

ANSWER:

(a) and (c). The integral test can be used because it is not hard to find the antiderivative of e^{-x}; the ratio test is often useful if exponential functions are involved.

COMMENT:

You might point out that this is a geometric series.

3. Which test will help you determine if the series converges or diverges?

$$\sum_{k=1}^{\infty} \frac{1}{k^3 + 1}$$

(a) Integral test
(b) Comparison test
(c) Ratio test

ANSWER:

(b). This series should be compared with $\sum_{k=1}^{\infty} \frac{1}{k^3}$.

COMMENT:

Ask the students if the series converges or diverges.

4. Which test will help you determine if the series converges or diverges?

$$\sum_{k=1}^{\infty} \frac{k}{(k+1)^2}$$

(a) Integral test
(b) Comparison test
(c) Ratio test

ANSWER:

(a) and (b). To use the comparison test, notice that $k/(k+1)^2 > k/(k+k)^2 = 1/4k$.

COMMENT:

You might want to point out that the substitution $w = x + 1$ is used to integrate $\int_1^{\infty} \frac{x}{(x+1)^2} dx$.

5. Which test will help you determine if the series converges or diverges?

$$\sum_{k=1}^{\infty} \frac{(-1)^{k-1}}{\sqrt{3k-1}}$$

(a) Alternating series test
(b) Integral test
(c) Ratio test

ANSWER:

(a). The alternating series test can be used since $a_k = 1/\sqrt{3k-1}$ is a positive sequence which decreases monotonically to 0 as $k \to \infty$.

COMMENT:

Follow-up Question. Does the series converge absolutely, converge, or diverge?
Answer. The series converges but does not converge absolutely.

6. Which test will help you determine if the series converges or diverges?

$$\sum_{k=1}^{\infty} \frac{(-1)^{k+1}5}{k}$$

(a) Alternating series test
(b) Absolute convergence test (that is, convergence of $\sum |a_n|$ implies convergence of $\sum a_n$)
(c) Ratio test

ANSWER:

(a). The alternating series test can be used since $a_k = 5/k$ is a positive sequence which decreases monotonically to 0 as $k \to \infty$.

COMMENT:

Follow-up Question. Does the series converge absolutely, converge, or diverge?
Answer. The series converges but does not converge absolutely.

7. Which test will help you determine if the series converges or diverges?

$$\sum_{k=1}^{\infty} \frac{(-1)^k 3^{k-1}}{k!}$$

    (a) Alternating series test

    (b) Absolute convergence test(that is, convergence of $\sum |a_n|$ implies convergence of $\sum a_n$)

    (c) Ratio test

        ANSWER:

        (a) and (c). The ratio test may be easier to use, but $3^{k-1}/k!$ is monotone for $k \geq 4$, since $3/k < 1$.

        COMMENT:

        Follow-up Question. Does the series converge absolutely, converge, or diverge?

        Answer. The series converges absolutely.

8. Which test can help you determine if the series converges or diverges?

$$\sum_{n=1}^{\infty} \frac{2000n}{2n^3 - 1}$$

    (a) Ratio test

    (b) Limit comparison test

    (c) Alternating series test

        ANSWER:

        (b). A limit comparison can be made with $\sum 1/n^2$.

        We are investigating the series $\sum_{n=1}^{\infty} a_n$ where $a_n = 2000n/(2n^3 - 1)$. With $b_n = 1/n^2$ we have

$$\lim_{n \to \infty} \frac{a_n}{b_n} = \lim_{n \to \infty} \frac{2000n/(2n^3 - 1)}{1/n^2} = \lim_{n \to \infty} \frac{2000n^3}{2n^3 - 1} = 1000.$$

    Since $\sum 1/n^2$ converges, so does $\sum a_n$.

        The ratio test does not apply because $\lim_{n \to \infty} |a_{n+1}|/|a_n| = 1$.

        The alternating series test does not apply to series whose terms do not alternate in sign.

9. Which test will help you determine if the series converges or diverges?

$$\sum_{k=1}^{\infty} \frac{(-1)^k 3}{k^2}$$

    (a) Alternating series test

    (b) Absolute convergence test (that is, convergence of $\sum |a_n|$ implies convergence of $\sum a_n$)

    (c) Ratio test

        ANSWER:

        (a) and (b). The absolute convergence test is a good choice since $\left| \frac{(-1)^k 3}{k^2} \right| = \frac{3}{k^2}$ and $\sum \frac{3}{k^2}$ converges.

        COMMENT:

        Follow-up Question. Does the series converge absolutely, converge, or diverge?

        Answer. The series converges absolutely.

10. In order to determine if the series converges or diverges, the comparison test can be used. Decide which series provides the best comparison.

$$\sum_{k=1}^{\infty} \frac{\sqrt{k+1}}{k^2 + 1}$$

(a) $\displaystyle\sum_{k=1}^{\infty} \frac{1}{k}$ (b) $\displaystyle\sum_{k=1}^{\infty} \frac{1}{k^{3/2}}$ (c) $\displaystyle\sum_{k=1}^{\infty} \frac{\sqrt{2k}}{k^2}$

    ANSWER:

    (c). The series in (a) diverges and is greater than the original series, so (a) provides no help. The series in (b) converges, but it is smaller than the original series for $k \geq 3$.

        COMMENT:

        Graphing $f(x) = \sqrt{x+1}/(x^2 + 1)$ and $g(x) = \sqrt{2x}/x^2$ on the same axes may help students compare the two series.

11. In order to determine if the series converges or diverges, the comparison test can be used. Decide which series provides the best comparison.

$$\sum_{k=2}^{\infty} \frac{\ln k}{k^2}$$

(a) $\displaystyle\sum_{k=2}^{\infty} \frac{1}{k}$ (b) $\displaystyle\sum_{k=2}^{\infty} \frac{1}{k^{3/2}}$ (c) $\displaystyle\sum_{k=2}^{\infty} \frac{1}{k^2}$

ANSWER:

(b). The series in (a) diverges and is greater than the original series, so (a) provides no help. The series in (c) converges, but for $k > 2$, its terms $1/k^2$ are smaller than the terms $\ln k/k^2$ of the original series.

COMMENT:

Graphing $f(x) = \ln x/x^2$ and $g(x) = 1/x^{3/2}$ on the same axes may help students compare the two series.

12. In order to determine if the series converges or diverges, the comparison test can be used. Decide which series provides the best comparison.

$$\sum_{k=1}^{\infty} \left(\frac{1}{k} - \frac{1}{k+1} \right)$$

(a) $\displaystyle\sum_{k=1}^{\infty} \frac{1}{k}$ (b) $\displaystyle\sum_{k=1}^{\infty} \left(-\frac{1}{k+1} \right)$ (c) $\displaystyle\sum_{k=1}^{\infty} \frac{1}{k^2}$

ANSWER:

(c). The original series can be rewritten as $\displaystyle\sum_{k=1}^{\infty} \frac{1}{k(k+1)}$.

COMMENT:

You might point out that this series is also known as a collapsing sum. By writing out the first few terms, the students should be able to identify the sum of this series.

13. In order to determine if the series converges or diverges, the comparison test can be used. Decide which series provides the best comparison.

$$\sum_{k=1}^{\infty} \frac{\cos k}{2^k}$$

(a) $\displaystyle\sum_{k=1}^{\infty} \cos k$ (b) $\displaystyle\sum_{k=1}^{\infty} \frac{(-1)^k}{2^k}$ (c) $\displaystyle\sum_{k=1}^{\infty} \frac{1}{2^k}$

ANSWER:

(c). Since $|\cos k| \leq 1$ for all k, we have $|\cos k/2^k| \leq 1/2^k$. We use the comparison test to prove absolute convergence and then the absolute convergence test to prove convergence of the original series.

COMMENT:

This question could be repeated with any bounded function $f(k)$ in place of $\cos k$.

14. Arrange the values of these two alternating series and two numbers in order from smallest to largest.

(I) $1 - \dfrac{1}{4} + \dfrac{1}{9} - \dfrac{1}{16} + \cdots + \dfrac{(-1)^n}{(n+1)^2} + \cdots$

(II) $1 - \dfrac{1}{7} + \dfrac{1}{49} - \dfrac{1}{343} + \cdots + \dfrac{(-1)^n}{7^n} + \cdots$

(III) $3/4$

(IV) $31/36$

(a) (I) (II) (III) (IV)

(b) (III) (I) (IV) (II)

(c) (II) (III) (I) (IV)

(d) (III) (I) (II) (IV)

(e) None of the first four answers is correct.

ANSWER:

(b). Both (I) and (II) are alternating series, $\sum (-1)^{n-1} a_n$, for which $0 < a_{n+1} < a_n$ and $\lim_{n \to \infty} a_n = 0$, so the sum of the series lies between successive partial sums of the series. Thus, taking the first two terms in series (I) gives $1 - 1/4 = 3/4$, so the sum of the series is greater than $3/4$. Taking one more term gives $3/4 + 1/9 = 31/36$. So, the sum of (I) is less than $31/36$. Series (II) is a geometric series with a sum of $1/(1 + 1/7) = 7/8$. Thus (III) $= 3/4 <$ (I) $< 31/36 =$ (IV) $< 7/8 =$ (II). The order is (III) $<$ (I) $<$ (IV) $<$ (II).

COMMENT:

Follow-up Question. How many terms do you need to add together in series (I) in order to obtain an accuracy to three decimal places?

Answer. We want to find n so that the sum of the first n terms, S_n, and the first $n+1$ terms, S_{n+1}, differ in magnitude by 0.0005, that is $|S_n - S_{n+1}| < 0.0005$. But $|S_n - S_{n+1}| = 1/(n+1)^2$ so we need $(n+1)^2 > 1/0.0005$, so n needs to be at least 44.

15. For each of the following situations, decide if $\displaystyle\sum_{n=1}^{\infty} c_n$ converges, diverges, or if one cannot tell without more information.

(a) $0 \leq c_n \leq \dfrac{1}{n}$ for all n. (b) $\dfrac{1}{n} \leq c_n$ for all n. (c) $0 \leq c_n \leq \dfrac{1}{n^2}$ for all n.

(d) $\dfrac{1}{n^2} \leq c_n$ for all n. (e) $\dfrac{1}{n^2} \leq c_n \leq \dfrac{1}{n}$ for all n.

ANSWER:

(a) Cannot tell without more information.

(b) Diverges.

(c) Converges.

(d) Cannot tell without more information.

(e) Cannot tell without more information.

COMMENT:

Follow-Up Question. For each case where you need more information, find an example of a series that converges and an example of a series that diverges both of which satisfy the given condition.

Answer. (a) $1/(n+1) < 1/n$ for all n, but using the integral test, the series $\sum 1/(n+1)$ diverges. The series $\sum 1/n^2$ converges by the integral test. (d) $1/n^2 < 1/(n^2 - 1/4) < 1/(n+1)$ for all n. The series $\sum 1/(n^2 - 1/4)$ converges by the integral test and the series $\sum 1/(n+1)$ diverges. (e) You can use the same examples as those given in the answer to (d).

16. (a) A student says: "The series $\displaystyle\sum_{n=1}^{\infty} \dfrac{1}{n}$ converges since $\lim_{n \to \infty} 1/n = 0$." What is wrong with this statement?

(b) A student says: "The series $\displaystyle\sum_{n=1}^{\infty} (-1)^n n^2$ converges by the alternating series test." What is wrong with this statement?

ANSWER:

(a) Knowing $\lim_{n \to \infty} a_n = 0$ does not tell us that $\sum a_n$ converges; the theorem tells us that if $\sum a_n$ converges, then $\lim_{n \to \infty} a_n = 0$. In fact, $\sum 1/n$ diverges. The theorem is most often useful when we know that $\lim_{n \to \infty} a_n \neq 0$, and then we can conclude that $\sum a_n$ does not converge.

(b) The alternating series test does not apply because n^2 is not a decreasing positive function of n.

COMMENT:

Follow-up Question. Explain how you know that the second series diverges.

Answer. Since $\lim_{n \to \infty} (-1)^n n^2 \neq 0$ we know that $\sum (-1)^n n^2$ diverges.

17. The limit comparison test can be used to determine whether the series converges. Decide which series to compare with.

$$\sum_{n=1}^{\infty} \frac{200n^2 - 100n - 1}{n^3 + n^2 + n + 1}$$

(a) $\sum 1/n^3$ (b) $\sum 1/n^2$ (c) $\sum 1/n$

ANSWER:

(c). We have

$$\lim_{n\to\infty} \frac{(200n^2 - 100n - 1)/(n^3 + n^2 + n + 1)}{1/n} = \lim_{n\to\infty} \frac{200n^3 - 100n^2 - n}{n^3 + n^2 + n + 1} = 200.$$

The limit comparison test applies. Since $\sum 1/n$ diverges, so does the series $\sum_{n=1}^{\infty}(200n^2 - 100n - 1)/(n^3 + n^2 + n + 1)$.

18. A student says that the series

$$\sum_{n=1}^{\infty} \frac{n}{1.05^n} = 0.95 + 1.81 + 2.59 + 3.29 + \cdots$$

diverges because the terms are getting larger. What is wrong with this statement?

ANSWER:

Only the first twenty terms of the series are getting larger. With $a_n = n/1.05^n$, we have $a_{21} = 7.54$, $a_{22} = 7.52$, $a_{23} = 7.49$ and for $n > 23$ the terms continue to decrease. We can investigate convergence of this series with the ratio test. We have

$$\lim_{n\to\infty} \frac{|a_{n+1}|}{|a_n|} = \lim_{n\to\infty} \frac{(n+1)/1.05^{n+1}}{n/1.05^n} = \lim_{n\to\infty} \frac{n+1}{n} \frac{1}{1.05} = \frac{1}{1.05} < 1,$$

which shows that the series converges.

Convergence of a series can never be determined by looking at only a finite number of terms of the series.

ConcepTests and Answers and Comments for Section 9.5

1. Determine which of the following series has the smallest radius of convergence.

(a) $\sum_{n=0}^{\infty} \frac{(x-10)^n}{6n+1}$ (b) $\sum_{n=0}^{\infty} \frac{(x-10)^n}{n+7}$

(c) $\sum_{n=0}^{\infty} \frac{2^n(x-10)^n}{n+1}$ (d) $\sum_{n=0}^{\infty} \frac{(x-10)^n}{\sqrt{n+1}}$

ANSWER:

(c). The ratio test gives $\lim_{n\to\infty} \left| \frac{2^{n+1}(x-10)^{n+1}/(n+2)}{2^n(x-10)^n/(n+1)} \right| = \lim_{n\to\infty} \left| 2(x-10)\frac{(1+1/n)}{(1+2/n)} \right| = 2|x-10|$. Thus the radius of convergence is $1/2$. The ratio test gives 1 as the radius of convergence of the other three series.

COMMENT:

You might point out that if the c_n in $c_n(x-a)^n$ are rational functions of n, the ratio c_{n+1}/c_n approaches 1 as n approaches infinity.

2. Determine which of the following series has the smallest radius of convergence.

(a) $\sum_{n=0}^{\infty} (-1)^n(n+2)(x-1)^n$ (b) $\sum_{n=0}^{\infty} \frac{(x-1)^n}{3^n}$

(c) $\sum_{n=0}^{\infty} \frac{(x-1)^n}{\sqrt{(n+1)!}}$ (d) $\sum_{n=0}^{\infty} 3^n(x-1)^n$

ANSWER:

(d). For this series the ratio test gives $\lim_{n\to\infty} \left| \frac{3^{n+1}(x-1)^{n+1}}{3^n(x-1)^n} \right| = \lim_{n\to\infty} |3(x-1)| = 3|x-1|$ so $R = 1/3$. The radius of convergence for the series in (a) is $R = 1$, for (b) $R = 3$, and for (c) $R = \infty$.

COMMENT:

You might point out that if the c_n in $c_n(x-a)^n$ are rational functions of n, the ratio c_{n+1}/c_n approaches 1 as n approaches infinity.

3. Determine the general term for the following series if the starting value is $n = 1$.

$$(x - 1) - \frac{(x - 1)^2}{2} + \frac{(x - 1)^3}{4} - \frac{(x - 1)^4}{8} + \frac{(x - 1)^5}{16} - \cdots.$$

(a) $\dfrac{(-1)^n (x - 1)^n}{2n}$

(b) $\dfrac{(-1)^{n+1} (x - 1)^n}{2^n}$

(c) $\dfrac{(-1)^n (x - 1)^n}{2^n}$

(d) $\dfrac{(-1)^{n+1} 2 (x - 1)^n}{2^n}$

ANSWER:

(d). Notice that the coefficient of $(x - 1)^n$ in the first term in (a) and (c) is negative, while the first term in (b) is $(x - 1)/2$.

COMMENT:

You could ask for a free response to this question instead giving multiple choices.

4. Determine the general term for the following series if the starting value is $n = 1$.

$$x^2 + \frac{x^4}{2} + \frac{x^6}{6} + \frac{x^8}{24} + \frac{x^{10}}{120} + \cdots$$

(a) $\dfrac{2x^n}{n!}$

(b) $\dfrac{x^{2n}}{n!}$

(c) $\dfrac{x^{n+2}}{(2n)!}$

(d) $\dfrac{(x^2)^n}{n!}$

(e) None of the above

ANSWER:

(b) and (d). The denominators, 1, 2, 6, 24, 120 follow the pattern of $n!$, whereas the numerators are powers of x^2.

COMMENT:

You could ask for a free response to this question instead giving multiple choices.

5. Determine the general term for the following series if the starting value is $n = 1$.

$$\frac{1}{2} + x + \frac{5x^2}{4} + \frac{7x^3}{5} + \frac{9x^4}{6} + \cdots$$

(a) $\dfrac{(2n - 1)x^{n-1}}{n + 1}$

(b) $\dfrac{(2n + 1)x^n}{n + 2}$

(c) $\dfrac{(2n - 1)x^n}{n}$

(d) $\dfrac{(2n + 1)x^{n-1}}{n + 1}$

(e) None of the above

ANSWER:

(a). The pattern is more clear if the second term is written as $\dfrac{3x}{3}$.

COMMENT:

You could ask for a free response to this question instead giving multiple choices.

6. Which of the following are power series?

   (a) $\sum_{n=0}^{\infty} \frac{1}{n+1}(x-5)^n$

   (b) $\sum_{n=7}^{\infty} \frac{1}{n}(x-5)^n$

   (c) $\sum_{n=0}^{\infty} \frac{1}{n+1}(n-5)^n$

   (d) x

   ANSWER:
   (a), (b), and (d) are power series.
   COMMENT:
   (c) is not a power series since it does not contain a variable.

7. The power series $\sum C_n x^n$ diverges at $x=7$ and converges at $x=-3$. At $x=-9$, the series is

   (a) Conditionally convergent
   (b) Absolutely convergent
   (c) Alternating
   (d) Divergent
   (e) Cannot be determined.

   ANSWER:
   (d). Since the series is centered at $x=0$ and diverges at $x=7$, the radius of convergence is $R \leq 7$. The series converges at $x=-3$, so $R \geq 3$. Since $|-9|=9 > R$, the series diverges at $x=-9$.

8. The power series $\sum C_n x^n$ diverges at $x=7$ and converges at $x=-3$. At $x=2.5$, the series is

   (a) Conditionally convergent
   (b) Absolutely convergent
   (c) Alternating
   (d) Divergent
   (e) Cannot be determined.

   ANSWER:
   (b). Since the series is centered at $x=0$ and diverges at $x=7$, the radius of convergence is $R \leq 7$. The series converges at $x=-3$, so $R \geq 3$. Since $|2.5|=2.5 < R$, the series is absolutely convergent at $x=2.5$.

9. The power series $\sum C_n x^n$ diverges at $x=7$ and converges at $x=-3$. At $x=-4$, the series is

   (a) Conditionally convergent
   (b) Absolutely convergent
   (c) Alternating
   (d) Divergent
   (e) Cannot be determined.

   ANSWER:
   (e). Since the series is centered at $x=0$ and diverges at $x=7$, the radius of convergence is $R \leq 7$. The series converges at $x=-3$, so $R \geq 3$. Since $|-4|=4$, which is between 3 and 7, we cannot tell whether the series converges at $x=-4$.

10. The power series $\sum C_n (x-5)^n$ converges at $x=-5$ and diverges at $x=-10$. At $x=11$, the series is

   (a) Conditionally convergent
   (b) Absolutely convergent
   (c) Alternating
   (d) Divergent
   (e) Cannot be determined.

   ANSWER:
   (b). Since the series is centered at $x=5$ and converges at $x=-5$, the radius of convergence is $R \geq 10$. The series diverges at $x=-10$, so $R \leq 15$. Since $|11-5|=6 < R$, the series converges absolutely at $x=11$.

11. The power series $\sum C_n (x - 5)^n$ converges at $x = -5$ and diverges at $x = -10$. At $x = 17$, the series is

    (a) Conditionally convergent
    (b) Absolutely convergent
    (c) Alternating
    (d) Divergent
    (e) Cannot be determined.

    ANSWER:

    (e). Since the series is centered at $x = 5$ and converges at $x = -5$, the radius of convergence is $R \geq 10$. The series diverges at $x = -10$, so $R \leq 15$. Since $|17 - 5| = 12$, which is between 10 and 15, we cannot tell whether the series converges at $x = 17$.

12. The power series $\sum C_n (x - 5)^n$ converges at $x = -5$ and diverges at $x = -10$. At $x = -13$, the series is

    (a) Conditionally convergent
    (b) Absolutely convergent
    (c) Alternating
    (d) Divergent
    (e) Cannot be determined.

    ANSWER:

    (d). Since the series is centered at $x = 5$ and converges at $x = -5$, the radius of convergence is $R \geq 10$. The series diverges at $x = -10$, so $R \leq 15$. Since $|-13 - 5| = 18 > R$, the series diverges at $x = -13$.

Chapter Ten

ConcepTests and Answers and Comments for Section 10.1

For Problems 1–10, suppose $P_2(x) = a + b(x-1) + c\dfrac{(x-1)^2}{2}$ is a Taylor polynomial of degree two about $x = 1$ for some function f. Give the signs of a, b, and c, if the graph of f is as shown. (Note: 0 is also a possible answer for a and b, but not for c.)

1.

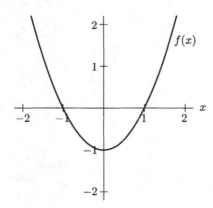

ANSWER:
$0, +, +$. The graph contains the point $(1, 0)$ and is increasing and concave up there.
COMMENT:
Follow-up Question. What would be the signs of a, b, and c if the Taylor series was about -1 instead of 1?
Answer. $0, -, +$, since the graph contains the point $(-1, 0)$ and is decreasing and concave up when $x = -1$.

2.

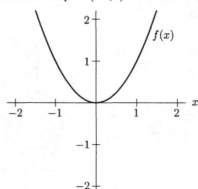

ANSWER:
$+, +, +$. At $x = 1$ the function is positive, increasing, and concave up.
COMMENT:
Follow-up Question. What would be the signs of a, b, and c if the Taylor series was about -1 instead of 1?
Answer. $+, -, +$. At $x = -1$ the function is positive, decreasing, and concave up.

3.

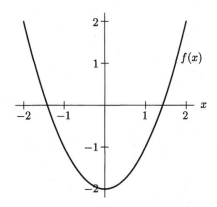

ANSWER:

$-, +, +$. At $x = 1$ the function is negative, increasing, and concave up.

COMMENT:

Follow-up Question. What would be the signs of a, b, and c if the Taylor series was about -1 instead of 1?

Answer. $-, -, +$. At $x = -1$ the function is negative, decreasing, and concave up.

4.

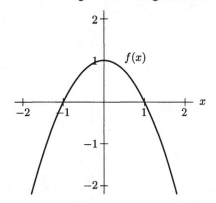

ANSWER:

$0, -, -$. The graph contains the point $(1, 0)$ and is decreasing and concave down there.

COMMENT:

Follow-up Question. What would be the signs of a, b, and c if the Taylor series was about -1 instead of 1?

Answer. $0, +, -$, since the graph contains the point $(-1, 0)$ and is increasing and concave down there.

5.

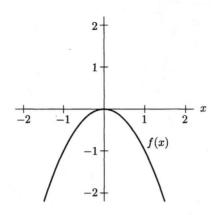

ANSWER:

$-, -, -$. At $x = 1$ the function is negative, decreasing, and concave down.

COMMENT:

Follow-up Question. What would be the signs of a, b, and c if the Taylor series was about -1 instead of 1?

Answer. $-, +, -$. At $x = -1$ the function is negative, increasing, and concave down.

6.

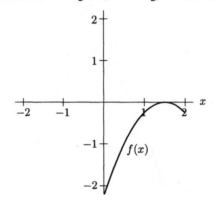

ANSWER:

$-, +, -$. At $x = 1$ the function is negative, increasing, and concave down.

COMMENT:

Follow-up Question. What would be the signs of a, b, and c if the Taylor series was about 1.5 instead of 1?

Answer. $0, 0, -$. The point $(1.5, 0)$ is on the graph. Also, when $x = 1.5$, the function $f(x)$ has a horizontal tangent and is concave down.

7.

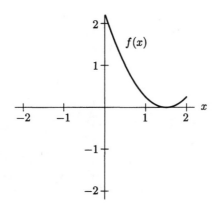

ANSWER:

$+$, $-$, $+$. At $x = 1$ the function is positive, decreasing, and concave up.

COMMENT:

Follow-up Question. What would be the signs of a, b, and c if the Taylor series was about 1.5 instead of 1?

Answer. 0, 0, $+$. The point $(1.5, 0)$ is on the graph. Also, when $x = 1.5$, the function $f(x)$ has a horizontal tangent and is concave up.

8.

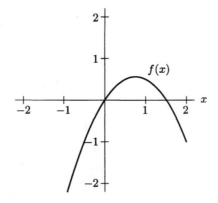

ANSWER:

$+$, $-$, $-$. At $x = 1$ the function is positive, decreasing, and concave down.

COMMENT:

Follow-up Question. What would be the signs of a, b, and c if the Taylor series was about 0 instead of 1?

Answer. 0, $+$, $-$. The point $(0, 0)$ is on the graph. Also, when $x = 0$, the function $f(x)$ is increasing and concave down.

9.

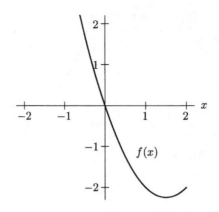

ANSWER:

$-$, $-$, $+$. At $x = 1$ the function is negative, decreasing, and concave up.

COMMENT:

Follow-up Question. What would be the signs of a, b, and c if the Taylor series was about 0 instead of 1?

Answer. 0, $-$, $+$. The point $(0, 0)$ is on the graph. Also, when $x = 0$, the function $f(x)$ is decreasing and concave up.

10.

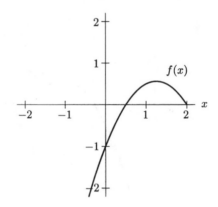

ANSWER:

$+$, $+$, $-$. At $x = 1$ the function is positive, increasing, and concave down.

COMMENT:

Follow-up Question. What would be the signs of a, b, and c if the Taylor series was about 0 instead of 1?

Answer. $-$, $+$, $-$. At $x = 0$, the function $f(x)$ is negative, increasing, and concave down.

11. Which of (a)–(d) is the graph of the addition of the two functions in Figures 10.1 and 10.2?

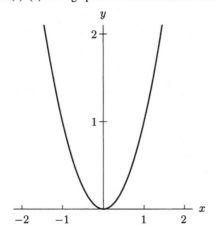

Figure 10.1

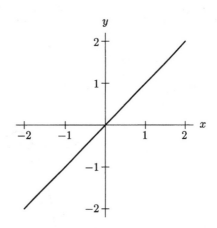

Figure 10.2

(a)
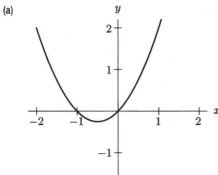

(b)
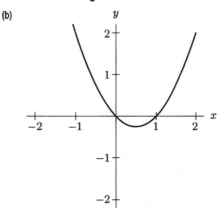

(c)
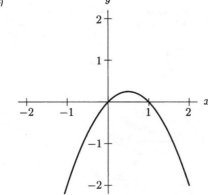

(d)
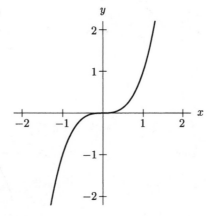

ANSWER:

(a). Both of the original functions are increasing for $x > 0$, so their sum must be also. This eliminates (b) and (c). At $x = -1$, one function has a value 1, while the other has the value -1, so their sum is 0.

COMMENT:

You could also check what happens to the addition of function values for several values of x.

Follow-up Question. What if you subtract the two functions? (You could ask them to do it both ways. They should realize that if both answers are graphed on the same set of axes, then one is a reflection across the x-axis of the other one.)

Answer. Answer (b) is the graph of the difference of the function in Figure 10.1 from the function in Figures 10.2. Answer (c) is the other difference between the two functions.

12. The graph in Figure 10.3 is generated by what combination of the graphs of the three functions f_1, f_2, and f_3?

(a) $f_1 + 0(f_2) + f_3$
(b) $f_1 - f_2 + f_3$
(c) $-f_1 + 0(f_2) + f_3$
(d) $-f_1 + 0(f_2) - f_3$
(e) $f_1 + 0(f_2) - f_3$
(f) $-f_1 + f_2 + f_3$

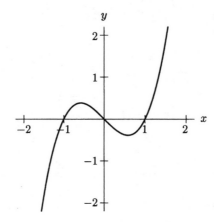

Figure 10.3

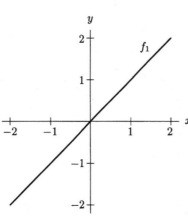

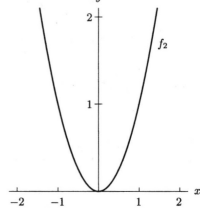

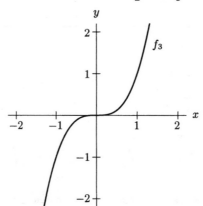

ANSWER:

(c). The graph describes an odd function. This eliminates (b) and (f). The point (1,0) is on the graph which eliminates (a) and (d). The functions is positive for $x > 1$ which eliminates (e).

COMMENT:

It may be worth asking which function, f_1, f_2, or f_3, is not like the graph in Figure 10.3 when the question is first presented to the class. This might encourage your students to think about an odd/even function argument.

13. The graph in Figure 10.4 is generated by what combination of the graphs of the three functions f_1, f_2, and f_3?

(a) $f_1 + f_2 + f_3$ (b) $f_1 - f_2 + f_3$ (c) $-f_1 + f_2 + f_3$

(d) $-f_1 + 0(f_2) - f_3$ (e) $f_1 + 0(f_2) + f_3$ (f) $f_1 + 0(f_2) - f_3$

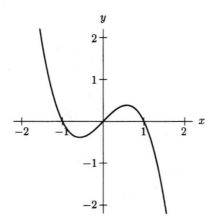

Figure 10.4

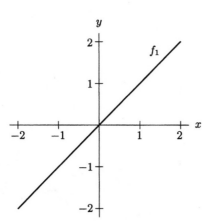

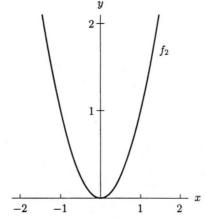

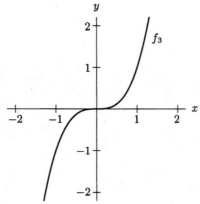

ANSWER:

(f). The point $(1,0)$ is on the graph which eliminates (a), (b), (c), (d), and (e).

COMMENT:

While the answer gives a quick solution, you may want to spend more time with geometric reasoning. For example, the fact that the graph in Figure 10.4 is an odd function eliminates (a), (b), and (c).

14. The graph in Figure 10.5 is generated by what combination of the graphs of the three functions f_1, f_2, and f_3?

(a) $f_1 + f_2 + f_3$

(b) $0(f_1) - f_2 + f_3$

(c) $-f_1 + 0(f_2) + f_3$

(d) $-f_1 + f_2 - f_3$

(e) $f_1 + 0(f_2) - f_3$

(f) $0(f_1) + f_2 + f_3$

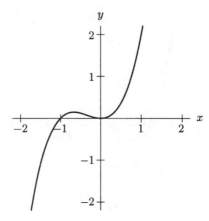

Figure 10.5

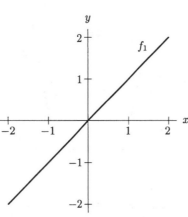

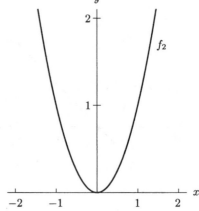

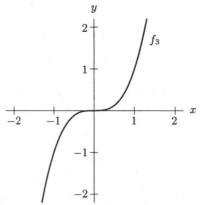

ANSWER:

(f). The function is not odd which eliminates (c) and (e). The point $(-1, 0)$ is on the graph which eliminates (a), (b), and (d).

COMMENT:

Notice that $f_3 - f_1$ has a curve that looks somewhat like Figure 10.5 when $x < 0$. However, they are quite different for $x > 0$.

ConcepTests and Answers and Comments for Section 10.2

1. Which of the following have Taylor series about the origin that do not converge for all values of x? Explain why.

   (a) e^{x-1}

   (b) $\dfrac{1}{x-4}$

   (c) $\arcsin(3x)$

   (d) $\sin(x^2)$

   ANSWER:

   (b) and (c). In (b) $\dfrac{1}{x-4} = -\dfrac{1}{4}\left(\dfrac{1}{1-\frac{x}{4}}\right) = -\dfrac{1}{4}\sum\left(\dfrac{x}{4}\right)^n$, which converges only for $-1 < x/4 < 1$. In (c) the

   Taylor series for $\arcsin x$ converges for $-1 < x < 1$, so the Taylor series for $\arcsin(3x)$ will converge for $-1 < 3x < 1$.

   COMMENT:

   You could ask students to come up with other examples.

ConcepTests and Answers and Comments for Section 10.3

1. Given that the radius of convergence of the Taylor series for $\ln(1-x)$ about $x = 0$ is 1, what is the radius of convergence for the following Taylor series about $x = 0$?

   (a) $\ln(4-x)$

   (b) $\ln(4+x)$

   (c) $\ln(1+4x^2)$

   ANSWER:

   (a) $\ln(4-x) = \ln(4(1-x/4)) = \ln(4) + \ln(1-x/4)$, so the Taylor series converges for $-1 < x/4 < 1$, or $-4 < x < 4$. The radius of convergence is 4.

   (b) $\ln(4+x) = \ln(4(1-(-x/4))) = \ln(4) + \ln(1-(-x/4))$, so the Taylor series converges for $-1 < -x/4 < 1$, or $4 > x > -4$. The radius convergence is 4.

   (c) $\ln(1+4x^2) = \ln(1-(-4x^2))$, so the Taylor series converges for $-1 < -4x^2 < 1$. This gives $x^2 < 1/4$, or $-1/2 < x < 1/2$. The radius of convergence is $1/2$.

   COMMENT:

   You could ask for the interval of convergence as you expand $\ln(4-x)$ about $x = 2$. (Here you need to write $4 - x = 4 - (x-2) - 2 = 2 - (x-2) = 2(1 - (x-2)/2).$)

2. Examine the Taylor series of the following functions to rank them in order of smallest to largest for small positive values of x.

   (a) $e^{-x/2}$

   (b) $\cos(\sqrt{x})$

   (c) $1 - \ln(1+x/2)$

   (d) $1 - \sin(x/2)$

   ANSWER:

   (a) $e^{-x/2} = 1 - x/2 + (x/2)^2/2 - \cdots = 1 - x/2 + x^2/8 - \cdots$

   (b) $\cos(\sqrt{x}) = 1 - x/2 + x^2/24 - \cdots$

   (c) $\ln(1+x/2) = 1 - x/2 + (x/2)^2 - \cdots = 1 - x/2 + x^2/4 + \cdots$

   (d) $1 - \sin(x/2) = 1 - x/2 + (x/2)^3/6 - \cdots = 1 - x/2 + x^3/48 - \cdots$

   Thus for small values of x the order is (d), (b), (a), (c).

   COMMENT:

   You could propose four other functions, where one is missing a first order term to point out the relative importance of the exponent over the coefficient.

3. Given that the Taylor series for $\cos x$ is $\displaystyle\sum_{n=0}^{\infty} \frac{(-1)^n x^{2n}}{(2n)!}$, then that of $(1/2)\cos(2x)$ is

   (a) $1 - x^2/2 + x^4/24 - \cdots$
   (b) $1/2 - x^2/2! + x^4/4! - \cdots$
   (c) $1/2 - x^2/4 + x^4/48 - \cdots$
   (d) $1/2 - x^2 + x^4/3 - \cdots$

   ANSWER:

   (d). The Taylor series for $(1/2)\cos(2x)$ is $1/2$ times the Taylor series for $\cos(2x)$, which is obtained from that of $\cos x$ by replacing x with $2x$.

   COMMENT:

   You could ask a similar question for another even function to point out what happens to the constant term is usually different from what happens to the higher order terms.

4. Given that the Taylor series for $\tan x = x + x^3/3 + 21x^5/120 + \cdots$, then that of $3\tan(x/3)$ is

   (a) $3x + x^3 + 21x^5/120 + \cdots$
   (b) $3x + x^3 + 21x^5/40 + \cdots$
   (c) $x + x^3/27 + 7x^5/3240 + \cdots$
   (d) $x + x^3/3 + 21x^5/120 + \cdots$

   ANSWER:

   (c). The Taylor series for $3\tan(x/3) = 3(x/3 + (x/3)^3/3 + 21(x/3)^5/120 + \cdots) = x + x^2/27 + 7x^5/3240 + \cdots$

   COMMENT:

   You could ask a similar question for another odd function or one that is neither even nor odd.

ConcepTests and Answers and Comments for Section 10.5

1. Figure 10.6 contains the graph of the first three terms of the Fourier series of which of the following functions?

   (a) $f(x) = \begin{cases} 0 & , \;\; -1 < x < 0 \\ 1 & , \;\; 0 < x < 1 \end{cases}$ and $f(x+2) = f(x)$

   (b) $f(x) = \begin{cases} -1 & , \;\; -1 < x < 0 \\ 1 & , \;\; 0 < x < 1 \end{cases}$ and $f(x+2) = f(x)$

   (c) $f(x) = |x|$ on $-1 < x < 1$ and $f(x+2) = f(x)$

   (d) $f(x) = \begin{cases} 1 + x & , \;\; -1 < x < 0 \\ 1 & , \;\; 0 < x < 1 \end{cases}$ and $f(x+2) = f(x)$

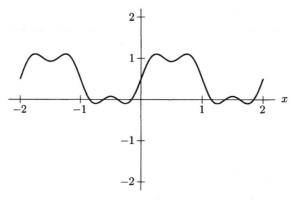

Figure 10.6

ANSWER:

(a). The graph describes a function that is neither even nor odd. So this eliminates (b) and (c). The Fourier series for (d) would have the value of 1 at $x = 0$, while the graph shows a value of $1/2$ there.

COMMENT:

Plotting the graphs in the four choices provides an efficient way to eliminate choices.

2. Figure 10.7 contains the graph of the first three terms of the Fourier series for which of the following functions?

  (a) $f(x) = 3(x/\pi)^3$ on $-\pi < x < \pi$ and $f(x + 2\pi) = f(x)$

  (b) $f(t) = |x|$ on $-\pi < x < \pi$ and $f(x + 2\pi) = f(x)$

  (c) $f(x) = \begin{cases} -3 & , \quad -\pi < x < 0 \\ 3 & , \quad 0 < x < \pi \end{cases}$ and $f(x + 2\pi) = f(x)$

  (d) $f(x) = \begin{cases} \pi + x & , \quad -\pi < x < 0 \\ \pi - x & , \quad 0 < x < \pi \end{cases}$ and $f(x + 2\pi) = f(x)$

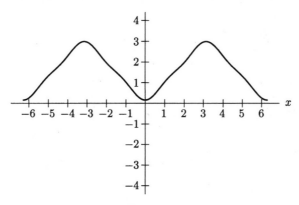

Figure 10.7

ANSWER:

(b). The graph describes an even function, which eliminates (a) and (c). The Fourier series for (d) would have values near π for x close to 0.

COMMENT:

Plotting the graphs in the four choices provides an efficient way to eliminate choices.

3. Figure 10.8 contains the graph of the first three non-zero terms of the Fourier series for which of the following functions?

(a) $f(x) = 3(x/\pi)^3$ on $-\pi < x < \pi$ and $f(x + 2\pi) = f(x)$

(b) $f(t) = |x|$ on $-\pi < x < \pi$ and $f(x + 2\pi) = f(x)$

(c) $f(x) = \begin{cases} -3 & , & -\pi < x < 0 \\ 3 & , & 0 < x < \pi \end{cases}$ and $f(x + 2\pi) = f(x)$

(d) $f(x) = \begin{cases} \pi + x & , & -\pi < x < 0 \\ \pi - x & , & 0 < x < \pi \end{cases}$ and $f(x + 2\pi) = f(x)$

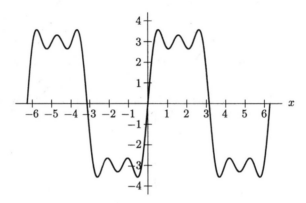

Figure 10.8

ANSWER:

(c). The graph describes an odd function which eliminates (b) and (d). The graph stays close to -3 for x between -3 and 0 which eliminates (a).

COMMENT:

Plotting the graphs in the four choices provides an efficient way to eliminate choices.

Chapter Eleven

ConcepTests and Answers and Comments for Section 11.1

For Problems 1–4, which of the following functions satisfy the given differential equation?

(a) e^{-x} (b) xe^{-x} (c) $3e^{-x}$ (d) e^{x} (e) xe^{x}

1. $y' = -y$
   ANSWER:
   (a) and (c)
   COMMENT:
   It may be useful to ask a similar question with functions such as e^{2x}, xe^{2x}, e^{-2x}, and xe^{-2x} with the differential equations $y' = 2y$, $y' = -2y$, $y'' = 4y$, $y'' = 4y' + 4y$, $y'' = -4y' + 4y$, and $y' = 2y + e^{2x}$.

2. $y'' - 2y' = -y$
   ANSWER:
   (d) and (e)
   COMMENT:
   See the Comment for Problem 1.

3. $y'' + 2y' = -y$
   ANSWER:
   (a), (b), and (c)
   COMMENT:
   See the Comment for Problem 1.

4. $y' + y = e^{-x}$
   ANSWER:
   (b)
   COMMENT:
   See the Comment for Problem 1.

For Problems 5–8, which of the following functions satisfy the given differential equation?

(a) $\sin x$ (b) $\cos x$ (c) $e^{x}\sin x$ (d) $e^{x}\cos x$ (e) e^{x}

5. $y'' = y$
   ANSWER:
   (e)
   COMMENT:
   It may be useful to ask a similar question with functions such as $\sin(2x)$, $\cos(2x)$, $e^{-x}\sin(2x)$, $e^{-x}\cos(2x)$, with the differential equations $y'' = 4y$, $y'' = -4y$, $y' + y = 2\cos(2x)$, $y'' + 2y' = -5y$.

6. $y'' = -y$
   ANSWER:
   (a) and (b)
   COMMENT:
   See the Comment for Problem 5.

7. $y'' - 2y' = -2y$
   ANSWER:
   (c) and (d)
   COMMENT:
   See the Comment for Problem 5.

8. $y' - y = e^{x}\cos x$
   ANSWER:
   (c)
   COMMENT:
   See the Comment for Problem 5.

9. Which of (a)–(d) is a graph of a function that is equal to its own derivative, i.e. $f'(x) = f(x)$ for all x?

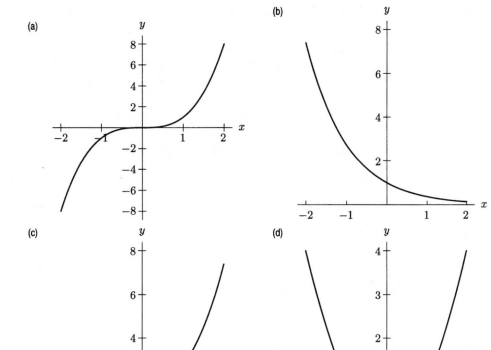

ANSWER:
(c). Because $f'(x) = f(x)$, the graph of f will be increasing when f is positive and decreasing when f is negative.
COMMENT:
This problem and the following one are repeats of earlier problems. They are meant as warm-up questions for Concept Questions 1 - 3 of the next section.

10. Which of (a)–(d) is a graph of a function that is equal to the negative of its own derivative, i.e. $f(x) = -f'(x)$?

(a)

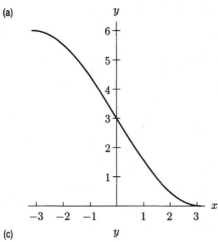

(b)

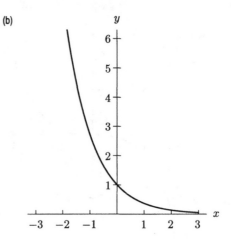

(c)

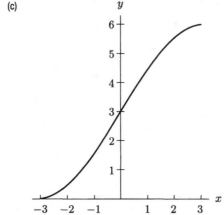

(d)

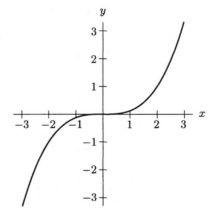

ANSWER:

(b). From $f(x) = -f'(x)$ we see that the graph will have negative slopes where it is positive and positive slopes where it is negative. This eliminates (c) and (d). Also, larger values of $f(x)$ require steeper slopes.

COMMENT:

Students may find other reasons for contradictions.

ConcepTests and Answers and Comments for Section 11.2

1. Which of the following graphs could be that of a solution to the differential equation $y' = -xy$?

(a)

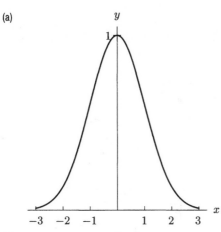

(b)

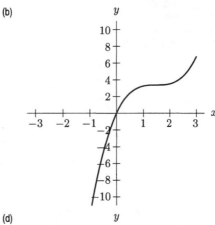

(c)

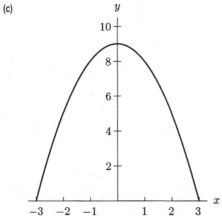

(d)

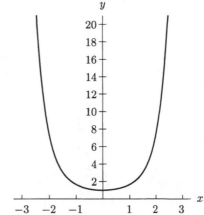

ANSWER:

(a). Notice that solutions have positive slopes if $x < 0$ and $y > 0$, and negative slopes if $x > 0$ and $y > 0$. Also the slope of the solution curve is zero if $x = 0$ or $y = 0$.

COMMENT:

You could have students eliminate other choices by finding regions of the graph that contradict the fact that its slope is given by $-xy$.

2. Which of the following graphs could be that of a solution to the differential equation $y' = -xy$?

   (a) (I)
   (b) (II)
   (c) (III)
   (d) (IV)
   (e) (I) and (IV)
   (f) (I), (II), and (IV)
   (g) (I) and (III)
   (h) (I), (II), and (III)
   (i) (I), (II), (III), and (IV)

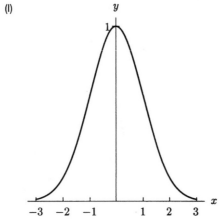

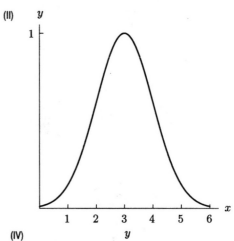

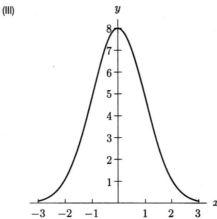

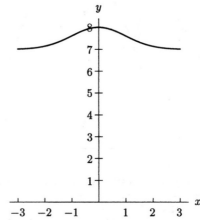

ANSWER:

   (g). From $y' = -xy$ we see that solutions have positive slopes for $x < 0$ and $y > 0$ and negative slopes for $x > 0$ and $y > 0$. This eliminates (II). The remaining three graphs have zero slope at $x = 0$ but the slope at the point $(3, 7)$ in (IV) should be -21, rather than almost zero.

COMMENT:

Students may find other reasons for eliminating choices.

3. Which of the following graphs could be that of a solution to the differential equation $y' = 3y/x$?

(a)

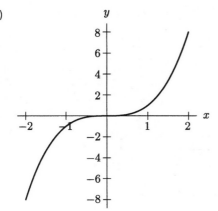

(b)

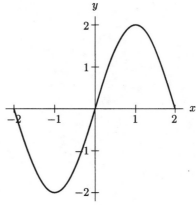

(c)

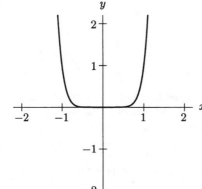

(d)

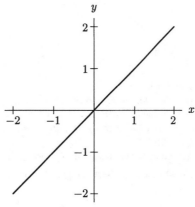

ANSWER:

(a). From $y' = 3y/x$, we see that solutions have positive slopes when x and y have the same sign. This eliminates (b). The slope when $y = x = 1$ must be 3 which eliminates (c) and (d).

COMMENT:

Students may find other reasons for eliminating choices.

4. Explain why the slope field in Figure 11.1 could only be that of exactly one of the following.

(a) $y' = xy$ (b) $y' = x/y$
(c) $y' = y/x$ (d) $y' = -xy$

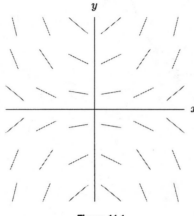

Figure 11.1

ANSWER:

(a). The slope field shows positive slopes in quadrants I and III, which eliminates (d). The slope field suggests horizontal tangents when $y = 0$, which eliminates (b). The slope field shows increasing slopes when x increases for a fixed y, which eliminates (c). The slope field is consistent with (a).

COMMENT:

You could repeat this question showing the slope field of $y' = -xy$.

5. Explain why the slope field in Figure 11.2 could only be that of exactly one of the following.

(a) $y' = xy$ (b) $y' = x/y$
(c) $y' = y/x$ (d) $y' = -xy$

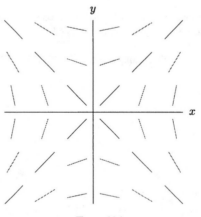

Figure 11.2

ANSWER:

(b). The slope field shows positive slopes in quadrants I and III, which eliminates (d). The slope field suggests horizontal tangents when $x = 0$, which eliminates (c). The slope field suggests vertical tangents when $y = 0$, which eliminates (a). The slope field is consistent with (b).

COMMENT:

You could repeat this question showing the slope field of $y' = y/x$.

ConcepTests and Answers and Comments for Section 11.3

1. For which of the following differential equations will Euler's method for approximating a solution which starts at $(0, 1)$ give an underestimate?

   (a) $y' = 2y$
   (b) $y' = -2y$
   (c) $y' = 2 - y$
   (d) $y' = y - 2$

   ANSWER:
   (a) and (b). For (a), differentiating gives $y'' = 2y' = 4y$, and for (b), we have $y'' = -2y' = 4y$. Thus solutions of these two equations that start at $(0, 1)$ are concave up, so tangent lines are below the curve, giving underestimates.

   COMMENT:
   You could also ask which differential equations lead to overestimates.

ConcepTests and Answers and Comments for Section 11.4

1. Which of the following are separable differential equations?

   (a) $y' = x(y^2 + 2)$
   (b) $y' = x^2 + xy$
   (c) $y' = e^{x+y}$
   (d) $y' = y \sin(2 + x)$
   (e) $y' = \ln x + \ln y$
   (f) $y' = x \cos(y^2)$
   (g) $y' = yx \arcsin x$

   ANSWER:
   (a), (c), (d), (f), and (g). The equations in (a), (d), (f), and (g) have the form $y' = g(x)f(y)$ or $y' = f(y)g(x)$, and so they are separable. (c) can be written as $y' = e^x e^y$, which is also separable. (b) can be written as $y' = x(x + y)$ and (e) as $y' = \ln(xy)$, neither of which are separable.

   COMMENT:
   Follow-up Question. Determine the solution of the separable equations.
   Answer. (f) does not have a simple antiderivative, and (g) will require integration by parts.

ConcepTests and Answers and Comments for Section 11.5

1. Which of the following slope fields indicate at least one equilibrium solution?

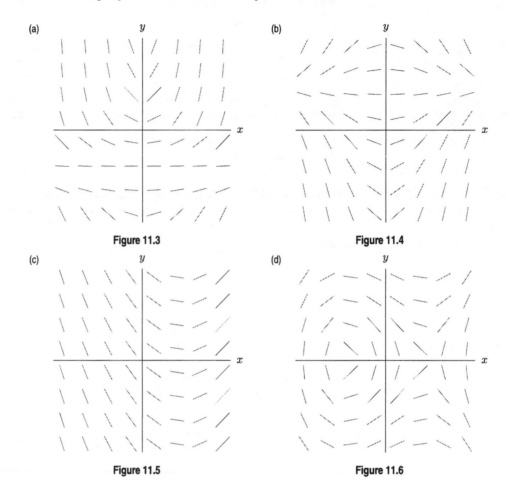

(a)

Figure 11.3

(b)

Figure 11.4

(c)

Figure 11.5

(d)

Figure 11.6

ANSWER:

(a) and (b). On a graph, an equilibrium solution is a horizontal line. Thus on a slope field we should look for short horizontal lines the same distance from the x-axis.

COMMENT:

This question may be repeated with other choices.

2. A yam is put in an oven where its temperature, H, is modeled by $\dfrac{dH}{dt} = -k(H - H_0)$, where t is time and k and H_0 are positive constants. Figure 11.7 shows four solutions of this differential equation.

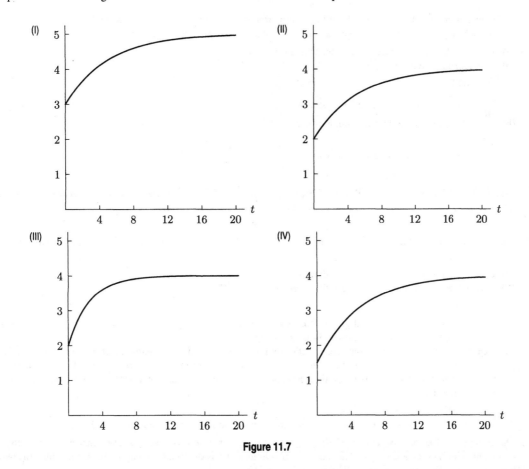

Figure 11.7

Which curve(s) correspond(s) to the

(a) Warmest oven?
(b) Lowest initial temperature?
(c) Same initial temperature?
(d) Largest value of k?

ANSWER:

(a) The warmest oven results in the largest final temperature, therefore (I).
(b) The lowest initial temperature is shown by (IV).
(c) (II) and (III) have the same initial temperature.
(d) Because $\dfrac{dH}{dt} = -k(H - H_0)$, and at $t = 0$ the values of $H - H_0$ are equal to -2 for the curves in (I), (II) and (III), the curve with the largest initial slope will correspond to the largest value of k. The curve (III) has the largest initial slope of these three curves. The value of $H - H_0$ at $t = 0$ for the curve (IV) is smaller than -2 and the initial slope is slightly less than that in curve (III), so the value 0f k for curve (IV) must be smaller than that of curve (III).

COMMENT:

You could point out that we can only make these comparisons because the scales on both axes are identical.

ConcepTests and Answers and Comments for Section 11.6 ▬▬▬▬▬▬▬

1. A chemical is produced at a rate which is directly proportional to both $(a - y)$ and $(t + b)$, and inversely proportional to $(y + c)$, where y is the amount of chemical at time t, and a, b, and c are positive constants.

   (a) Write the differential equation that describes this process.
   (b) Give all equilibrium solutions.

   ANSWER:

   (a) $\dfrac{dy}{dt} = \dfrac{k(a - y)(t + b)}{y + c}$, where k is a positive constant.
   (b) There is only one equilibrium solution, $y = a$.

   COMMENT:
   You could ask students to explain why $k > 0$.

ConcepTests and Answers and Comments for Section 11.7 ▬▬▬▬▬▬▬

1. Consider solutions to a differential equation modeling exponential growth and solutions to a differential equation modeling logistic growth.

   (a) Give at least two similarities between these solutions.
   (b) Give at least two differences between these solutions.

   ANSWER:

   (a) Solutions modeling exponential growth are increasing and concave up, which is also true for small values of the dependent variable for logistic growth. Both are always positive.
   (b) Exponential solutions are unbounded, logistic solutions are bounded. Exponential solutions are concave up everywhere, logistic solutions are concave down as they approach the carrying capacity from below. Exponential solutions have no inflection points, logistic solutions have an inflection point if their initial value is positive and less than half the carrying capacity.

   COMMENT:
   To illustrate, you could graph solutions to two specific equations with the same initial conditions.

2. The Gompertz differential equation, $dP/dt = -kP\ln(P/L)$, where k and L are positive constants, is used to model tumors in animals. List at least two characteristics of solutions of this differential equation that are common to solutions of the logistic differential equation, $dP/dt = kP(1 - P/L)$.

   ANSWER:
   Both have an equilibrium solution at $P = L$, and are increasing for $0 < P < L$. For the Gompertz equation,

   $$\frac{d^2 P}{dt^2} = -k\left(\frac{dP}{dt}\ln(P/L) + \frac{dP}{dt}\right) = -k(\ln(P/L) + 1)\frac{dP}{dt},$$

   so solutions to both equations are concave down for values of P just below the equilibrium solution at L.

   COMMENT:
   To illustrate, you could graph solutions to two specific equations with the same initial conditions.

ConcepTests and Answers and Comments for Section 11.8

1. Figure 11.8 shows a phase trajectory for a SI-phase plane modeling an epidemic. From this figure, estimate

   (a) The size of this population.
   (b) The maximum number infected, and the number who have never been infected at this time.
   (c) The number that never were infected during this epidemic.

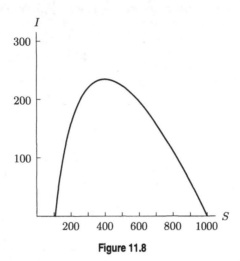

Figure 11.8

ANSWER:

(a) 1000

(b) 230 and 400. Some have already recovered by this time, so these do not add to 1000.

(c) 100

COMMENT:

In estimating numbers from a graph, a range of answers is possible for choice part (b).

ConcepTests and Answers and Comments for Section 11.9

1. Graph nullclines and determine all equilibrium points for the following differential equations.

   (a) $dx/dt = (x-1)(y-4)$, $dy/dt = (x-2)(y-3)$
   (b) $dx/dt = (x-4)(x-y-1)$, $dy/dt = y(2x-y-4)$

   ANSWER:

   (a) The nullclines where trajectories are horizontal are $x = 2$, and $y = 3$, those where the trajectories are vertical are $x = 1$ and $y = 4$. Thus the equilibrium points are $(1,3)$ and $(2,4)$. This is shown in Figure 11.9.

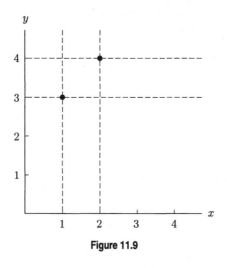

| Figure 11.9 | Figure 11.10 |

   (b) The nullclines where trajectories are horizontal are $y = 0$ and $y = 2x - 4$, those where the trajectories are vertical are $x = 4$ and $y = x - 1$. Thus the equilibrium points are $(1, 0)$, $(4, 0)$, $(3, 2)$, and $(4, 4)$. This is shown in Figure 11.10.

   COMMENT:

   Using other products of functions of x and y allows you to ask the same question for other differential equations.

ConcepTests and Answers and Comments for Section 11.10

1. The motion of a mass on a spring is governed by the initial value problem $y'' + py = 0$, $y(0) = A$, $y'(0) = B$, where p, A, and B are positive constants.

   (a) What is the effect of increasing p?
   (b) What is the effect of increasing B?
   (c) What is the effect of increasing A?
   (d) What is the effect of making B a negative number?
   (e) What is the effect of making A a negative number?

   ANSWER:

   (a) Increasing p decreases the period of the oscillations.
   (b) Increasing B for $B > 0$ increases the initial velocity, and therefore the amplitude of the oscillations if $A = 0$.
   (c) Increasing A for $A > 0$ increases the initial displacement, and therefore the amplitude of the oscillations, if $B = 0$.
   (d) If B is a negative number and A positive, the initial motion will be toward the equilibrium position. If B and A are both negative, the initial motion will be away from the equilibrium point.
   (e) If A is a negative number, the initial displacement is on the opposite side of the equilibrium position than when it is positive.

   COMMENT:

   You could repeat this questions having $A < 0$, $B > 0$; then $A > 0$, $B < 0$; and finally, $A < 0$, $B < 0$.

2. Which of the following graphs is that of a function that satisfies the relationship $f''(x) = -f(x)$?

(a)

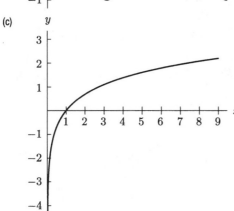

(b)

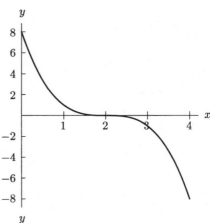

(c)

(d)

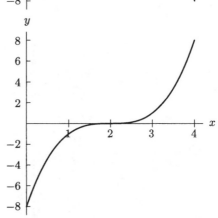

ANSWER:

(a). Because $f''(x) = -f(x)$, solution curves will be concave down when the solution is positive, and concave up where the solution is negative.

COMMENT:

You could also ask what $f''(x) = -f(x)$ implies about the location of the inflection points.

3. Which of the following graphs is that of a function that satisfies the relationship $f''(x) = -f(x)$?

   (a) (I), (II), and (III)
   (b) (I) and (II)
   (c) (II) and (III)
   (d) (I) and (III)
   (e) (I)
   (f) (II)
   (g) (III)
   (h) None of these

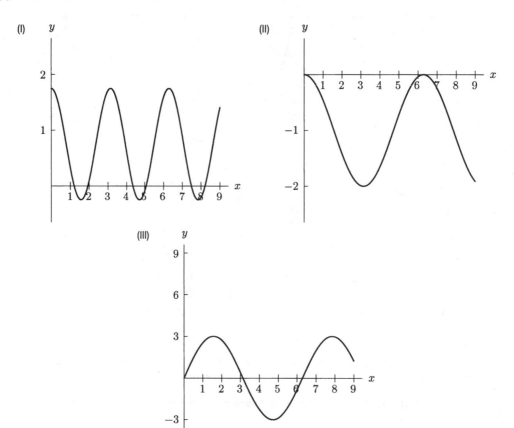

ANSWER:

(g). Because $f''(x) = -f(x)$, solution curves will be concave down when the solution is positive and concave up when the solution is negative.

COMMENT:

You might point out that while functions that satisfy $f''(x) = -f(x)$ are periodic, the converse is not true.

ConcepTests and Answers and Comments for Section 11.11

1. Match the following differential equations with the solutions shown in Figure 11.11.

   (a) $x'' + x = 0$
   (b) $x'' - x = 0$
   (c) $x'' + x' + 5x = 0$
   (d) $x'' - x' + 5x = 0$

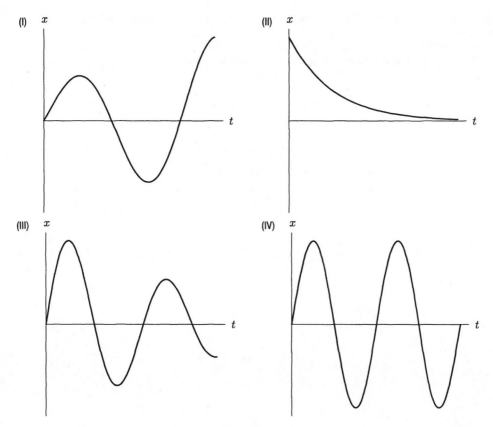

Figure 11.11

ANSWER:

(a) corresponds to (IV) as it describes undamped motion,
(b) corresponds to (II) as there are no oscillations,
(c) corresponds to (III) as it describes underdamped motion,
(d) corresponds to (I) as the motion is oscillatory, but with increasing amplitude.

COMMENT:

You could have students provide the analytical solution to these differential equations.

2. The motion of a mass on a spring is governed by $4x'' + 4bx' + 9x = 0$ where b is a positive constant.

   (a) For what values of b is the motion underdamped?
   (b) For what values of b is the motion overdamped?
   (c) For what values of b is the motion critically damped?

   ANSWER:
   The characteristic equation is $4r^2 + 4br + 9 = 0$, with the solutions

$$r = \frac{-4b \pm \sqrt{16b^2 - 4 \cdot 4 \cdot 9}}{2 \cdot 4} = \frac{-b \pm \sqrt{b^2 - 9}}{2}.$$

Thus motion is underdamped if $b^2 - 9 < 0$, i.e. $0 < b < 3$. The motion is overdamped if $b^2 - 9 > 0$, i.e. $b > 3$, and critically damped if $b = 3$.

   COMMENT:
   Follow-up Question. What would happen if b was a negative number?

Chapter Twelve

ConcepTests and Answers and Comments for Section 12.1 ───────

1. Look at the room you are in. Put a coordinate system in the room so that one corner is the origin. Find equations of the planes that describe all walls, the floor and the ceiling.

   ANSWER:

   The answer depends on the room, the choice of coordinate axes, and unit of distance.

   COMMENT:

   There will probably be a discussion on which corner to pick for the origin and what directions are x and y. As a follow-up question, ask for coordinates of certain points inside and outside of the classroom.

2. Which of the points $A = (3, 0, 3)$, $B = (0, 4, 2)$, $C = (2, 4, 1)$, and $D = (2, 3, 4)$ lies closest to

   (a) the xy-plane?   (b) the origin?
   (c) the y-axis?   (d) the point $(1, 2, 3)$?

   ANSWER:

   C is closest to the xy-plane, since the distance to the xy-plane is $|z|$.

   A is closest to the origin, since the distance to the origin is $\sqrt{x^2 + y^2 + z^2}$.

   B is closest to the y-axis, since the distance to the y-axis is $\sqrt{x^2 + z^2}$.

   D is closest to the point $(1, 2, 3)$, since the distance to $(1, 2, 3)$ is $\sqrt{(x - 1)^2 + (y - 2)^2 + (z - 3)^2}$.

   COMMENT:

   Ask the students to plot the four points.

3. In words, describe the surface given by the equation $\sqrt{(x - 1)^2 + (y + 2)^2 + (z + 5)^2} = 5$.

   ANSWER:

   A sphere of radius 5 centered at $(1, -2, -5)$.

   COMMENT:

   It is helpful to students to be able to recognize the equation for a sphere, a cylinder, a cone, and a paraboloid.

4. Sphere A is centered at the origin and the point $(0, 0, 3)$ lies on it. Sphere B is given by the equation $x^2 + y^2 + z^2 = 3$. Which of the following is true?

   (a) A encloses B   (b) B encloses A
   (c) A and B are equal   (d) None of the above

   ANSWER:

   (a) Both spheres are centered at the origin. Sphere A has radius 3, sphere B has radius $\sqrt{3}$. Thus, A encloses B.

   COMMENT:

   Follow-up Question. How would you change the equations of the spheres so that each of the other three choices happens?

ConcepTests and Answers and Comments for Section 12.2 ───────

1. Let $h(x, t) = 3 + 3 \sin(\frac{\pi}{10} x) \cos(2\pi t)$ be the distance above the ground (in feet) of a jump rope x feet from one end after t seconds. The two people turning the rope stand 10 feet apart. Then $h(x, \frac{1}{4})$ is

   (a) Concave up   (b) Concave down
   (c) Flat   (d) Changes concavity in the middle

   ANSWER:

   (c) At $t = \frac{1}{4}$ the cosine is zero. Therefore, $h(x, \frac{1}{4}) = 3$ is constant, so its graph is a straight line.

   COMMENT:

   Follow-up Question. Find a value for t at which the cross section is concave down.

2. The object in 3-space described by $x = 2$ is

   (a) A point   (b) A line
   (c) A plane   (d) Undefined.

   ANSWER:

   (c), a plane. While x is fixed at 2, y and z can vary freely.

   COMMENT:

   If all three variables do not appear in a problem, students sometimes forget that the missing variables are important.

3. The set of points whose distance from the z-axis equals the distance from the xy-plane describes a

(a) Plane   (b) Cylinder   (c) Sphere

(d) Cone   (e) Double cone (two cones joined at their vertices)

ANSWER:

(e), double cone. Fix a value for z, for example, $z = a$, and draw the set of points in the plane $z = a$ that are equidistant from the z-axis and the xy-plane. It is the set of points at distance $|a|$ from the z-axis, namely a circle of radius $|a|$ in the plane $z = a$. Putting all these circles together, we get two cones, with vertex at the origin, one above the xy-plane and one below it.

COMMENT:

Note that you also have to consider negative values of z. Students have a hard time visualizing surfaces in 3 dimensions.

4. Which of the following objects cannot be obtained as the graph of a function of two variables with a single formula?

(a) Paraboloid

(b) Plane

(c) Cylinder

(d) Sphere

(e) Line

(f) Parabolic cylinder

ANSWER:

(c), cylinder; (d), sphere; (e), line.

COMMENT:

Follow-up Question. Give examples of functions whose graphs are the other surfaces.

ConcepTests and Answers and Comments for Section 12.3

1. Figure 12.1 shows contours of a function. You stand on the graph of the function at the point $(-1, -1, 2)$ and start walking parallel to the x-axis. Describe how you move and what you see when you look right and left.

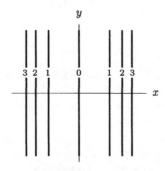

Figure 12.1

ANSWER:

You start to the left of the y-axis on the contour labeled 2. At first you go downhill, and the slope is steep. Then it flattens out. You reach a lowest point right on the y axis and start going uphill; the slope is increasing as you go. To your left and right the terrain is doing the exact same thing; you are walking across a valley that extends to your right and left.

COMMENT:

Follow-up Question. What happens if you walk in a different direction?

2. Which of the following is true? If false, find a counterexample.

   (a) The values of contour lines are always 1 unit apart.
   (b) Any contour diagram that consists of parallel lines comes from a plane.
   (c) The contour diagram of any plane consists of parallel lines.
   (d) Contour lines can never cross.
   (e) The closer the contours, the steeper the graph of the function.

   ANSWER:

   (c) and (e) are true. For (a), Figure 12.40 in the text, a contour diagram of the corn production function is a counterexample. For (b), the parabolic cylinder, $z = x^2$, has parallel line contours. For (d), contour lines of the same value can cross, for example, the contour 0 in $z = xy$.

   COMMENT:

   Follow-up Question. How would you have to change the statements to make them true (if possible)?

3. Which of the following terms best describes the origin in the contour diagram in Figure 12.2?

   (a) A mountain pass   (b) A dry river bed
   (c) A mountain top   (d) A ridge

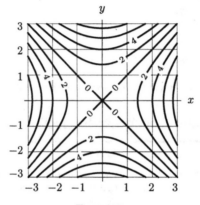

Figure 12.2

   ANSWER:
   A dry river bed.
   COMMENT:
   Follow-up Question. Imagine that you are walking on the graph of the function. Describe what you see and how you move.

4. Draw a contour diagram for the surface $z = 1 - x - y$. Choose a point and determine a path from that point on which

   (a) The altitude of the surface remains constant.
   (b) The altitude increases most quickly.

   ANSWER:

   To maintain a constant altitude, we move along a contour. Since the graph of this function is a plane, to find the path on which the altitude increases most quickly, we move perpendicular to the level curves.

   COMMENT:

   Choose other paths on the plane and describe them.

5. Which of the following surfaces do not have circles as contours?

   (a) A parabolic cylinder
   (b) A cone
   (c) A paraboloid
   (d) A hemisphere

   ANSWER:

   (a), a parabolic cylinder has straight line contours.

   COMMENT:

   Follow-up Question. How can you change the equation of the paraboloid $z = x^2 + y^2$ to make the contours ellipses?

ConcepTests and Answers and Comments for Section 12.4

1. Which of the following planes are parallel?

   (a) $z = -2 - 2x - 4y$
   (b) $z = -1 - x - 2y$
   (c) $(z - 1) = -2 - 2(x - 1) - 4(y - 1)$
   (d) $z = 2 + 2x + 4y$

   ANSWER:
   (a) and (c) are parallel.
   COMMENT:
   Follow-up Question. How could we change the other two so that they are parallel?

2. True or False? If false give a counterexample.

   (a) Any three points in 3-space determine a unique plane.
   (b) Any two distinct lines in 3-space determine a unique plane.
   (c) If the graph of $z = f(x, y)$ is a plane, then each cross section is a line.

   ANSWER:

   (a) False. If the three points lie on a line, there are many planes containing them.
   (b) False. If two lines don't intersect and are not parallel, then there is no plane that contains them both.
   (c) True.

   COMMENT:
   Ask what else must be assumed about the points in (a) and the lines in (b) to make the statements true.

3. Find as many conditions as possible that determine a unique plane. What information is needed to describe a unique plane?

   ANSWER:

   (a) Three points that are not collinear.
   (b) Two lines that intersect and are not identical.
   (c) Two parallel lines that are not identical.
   (d) A line and a point that does not lie on the line.
   (e) Other answers are possible.

   COMMENT:
   There are many different ways to describe a unique plane. Once we know vectors there are even more. This question can also be used at the end of Chapter 13.

4. Describe all the ways that three different planes can be positioned in 3-space so that they do not share any common points. For example: The three planes can all be parallel.

   ANSWER:

   (a) Two planes are parallel, and the third plane intersects each of the first two.
   (b) Two planes intersect in a line, and the third plane is parallel to that line, but not parallel to either of the first two planes. The three planes enclose an infinite triangular prism.

   COMMENT:
   It may take some time for students to come up with the second solution without any prompting. Playing around with sheets of paper may help to visualize the planes.

ConcepTests and Answers and Comments for Section 12.5 ━━━━━

1. Level surfaces of the function $f(x, y, z) = (x^2 + y^2)^{-1/2}$ are

   (a) Circles centered at the origin.
   (b) Spheres centered at the origin.
   (c) Cylinders centered around the z-axis.
   (d) Upper halves of spheres centered at the origin.

   ANSWER:

   (c). Cylinders centered around the z-axis.

   COMMENT:

   Follow-up Question. How do different level surfaces compare?

2. Describe in words the level surfaces of $V(x, y, z) = \ln(x^2 + y^2 + z^2)$.

   ANSWER:

   If $\ln(x^2 + y^2 + z^2) = C$, then $x^2 + y^2 + z^2 = e^C$. Level surfaces are spheres of radius $e^{C/2}$ centered at the origin.

   COMMENT:

   Follow-up Question. As C increases, what happens to the level surfaces?

3. True or false? If false, give a counterexample.

   (a) Any level surface of a function of 3 variables can be thought of as a surface in 3-space.
   (b) Any surface that is a graph of a 2-variable function $z = f(x, y)$ can be thought of as a level surface of a function of 3 variables.
   (c) Any level surface of a function of 3 variables can be thought of as the graph of a function $z = f(x, y)$.

   ANSWER:

   (a) False. The level surface of $f(x, y, z) = x^2 + y^2 + z^2$ given by $f(x, y, z) = 0$ is a single point, the origin $(0, 0, 0)$. The level surface of $g(x, y, z) = (x - y)^2 + z^2$ given by $g(x, y, z) = 0$ is the line $y = x$ in the xy-plane.
   (b) True
   (c) False, not every level surface is the graph of a function of two variables. For example: $V(x, y, z) = x^2 + y^2 + z^2$ has spheres as its level surfaces, which are not the graphs of functions of x and y.

   COMMENT:

   (b) and (c) are converses of each other. Only one is true. The implicit function theorem, available on the textbook website, sheds light on (a).

ConcepTests and Answers and Comments for Section 12.6 ━━━━━

In Problems 1–3, construct a function $f(x, y)$ with the given property.

1. Not continuous along the line $x = 2$; continuous everywhere else.

   ANSWER:

   One possible answer is $f(x, y) = \begin{cases} 0 & \text{if } x < 2 \\ 1 & \text{if } x \geq 2. \end{cases}$

2. Not continuous at the point $(2, 0)$; continuous everywhere else.

   ANSWER:

   One possible answer if $f(x, y) = 1/((x - 2)^2 + y^2)$.

3. Not continuous along the curve $x^2 + y^2 = 1$; continuous everywhere else.

   ANSWER:

   One possible answer is $f(x, y) = 1/(x^2 + y^2 - 1)$.

Chapter Thirteen

ConcepTests and Answers and Comments for Section 13.1

1. Consider the vector $\vec{v} = -\vec{i} + 2\vec{j} + 7\vec{k}$. Find a vector that

   (a) Is parallel but not equal to $\vec{v}$.
   (b) Points in the opposite direction of $\vec{v}$.
   (c) Has unit length and is parallel to $\vec{v}$.

   ANSWER:

   (a) $2\vec{v} = -2\vec{i} + 4\vec{j} + 14\vec{k}$
   (b) $-\vec{v} = \vec{i} - 2\vec{j} - 7\vec{k}$
   (c) $\vec{v}/\|\vec{v}\| = \frac{1}{\sqrt{54}}(-\vec{i} + 2\vec{j} + 7\vec{k})$

   Other answers are possible.

   COMMENT:
   Follow-up Question. How many vectors can you find in each case?

2. What would you get if you drew all possible unit vectors in 3-space with tails at the origin?

   ANSWER:
   A sphere of radius one.

   COMMENT:
   Follow-up Question. Is it just the surface of the sphere or is it a solid ball?

3. What object in three space is traced by the tips of all vectors starting at the origin that are of the form

   (a) $\vec{i} + \vec{j} + b\vec{k}$, where b is any real number.
   (b) $\vec{i} + a\vec{j} + b\vec{k}$, where a and b are any real numbers.

   ANSWER:

   (a) A vertical line, parallel to the z-axis.
   (b) The plane $x = 1$, parallel to the yz-plane.

   COMMENT:
   This problem provides valuable practice in visualizing objects in three space.

4. Describe the object created by all scalar multiples of $\vec{v} = \vec{i} + \vec{j}$ with tail at the point $(0, 0, 1)$.

   ANSWER:
   This describes a line through the point $(0, 0, 1)$ parallel to the vector $\vec{v} = \vec{i} + \vec{j}$. The line lies in the plane $z = 1$ and is parallel to the xy-plane.

   COMMENT:
   Problems like this help students visualize lines in three space. This particular way of describing a line comes up in the section on parameterizations.

5. Decide if each of the following statements is true or false:

   (a) The length of the sum of two vectors is always strictly larger than the sum of the lengths of the two vectors.
   (b) $\|\vec{v}\| = |v_1| + |v_2| + |v_3|$, where $\vec{v} = v_1\vec{i} + v_2\vec{j} + v_3\vec{k}$.
   (c) $\pm\vec{i}$, $\pm\vec{j}$, $\pm\vec{k}$ are the only unit vectors.
   (d) $\vec{v}$ and $\vec{w}$ are parallel if $\vec{v} = \lambda\vec{w}$ for some scalar λ.
   (e) Any two parallel vectors point in the same direction.
   (f) Any two points determine a unique displacement vector.
   (g) $2\vec{v}$ has twice the magnitude as $\vec{v}$.

   ANSWER:

   (a) False
   (b) False
   (c) False
   (d) True
   (e) False
   (f) False
   (g) True

   COMMENT:
   Follow-up Question. Find counterexamples for the false statements.

ConcepTests and Answers and Comments for Section 13.2

1. Decide if each of the following quantities is a vector or a scalar.

   (a) Velocity
   (b) Speed
   (c) Force
   (d) Area
   (e) Acceleration
   (f) Volume

   ANSWER:

   (a) Vector
   (b) Scalar
   (c) Vector
   (d) Scalar
   (e) Vector
   (f) Scalar

   COMMENT:
   Follow-up Question. Can you think of any other quantities that are vectors?

2. A plane wants to fly due south. There is a strong wind from the west. In what direction should the pilot point the plane to stay on course?

   (a) South
   (b) East
   (c) West
   (d) Between south and east
   (e) Between south and west

   ANSWER:
   (e), between south and west.
   COMMENT:
   Follow-up Question. What shape is the path of an object that is dropped from the plane?

3. A car drives around an elliptical track, always going as fast as it can (without flying off the track). Compare the speed and the velocity of the car at the given points in Figure 13.1.

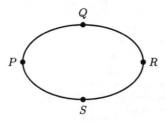

Figure 13.1

ANSWER:
The speeds are the same at P and R, but the velocity vectors point in opposite directions, tangent to the race track. The same is true of Q and S.
COMMENT:
Follow-up Question. What do the acceleration vectors look like at these points?

ConcepTests and Answers and Comments for Section 13.3

1. View the second and hour hands of a clock as vectors $\vec{s}$ and $\vec{h}$ in the plane. Describe $\vec{s} \cdot \vec{h}$ over the course of 2 minutes beginning at noon. If you increase the length of the second hand, what happens to the dot products?

    ANSWER:

    Over the course of two minutes, the hour hand moves very little, while the second hand makes two complete revolutions. The dot product starts out at a maximum with both vectors pointing at 12. As $\vec{s}$ moves, the dot product decreases until it is zero when the second hand is near 3, it keeps decreasing and becomes negative until it reaches a minimum when $\vec{s}$ points approximately toward 6. After that, the dot product increases, reaches zero when $\vec{s}$ points approximately at 9 and keeps increasing until it reaches a maximum when $\vec{s}$ points at about 12. During the second minute the dot product repeats the same pattern with a slight variation due to the small movement of the hour hand.

    If the length of $\vec{s}$ is increased, the dot product goes through the same sign changes and has the same zeros but reaches a larger maximum value and a lower (negative) minimum.

    COMMENT:

    This is a good place to reinforce that the dot product depends on both the magnitude and the angle between the vectors.

2. Which of the following planes are parallel to one another? Which are perpendicular to one another?

    (a) $2x + 3y - 5z = 4$ (b) $2x - 3y - z = 3$
    (c) $4x + 6y - 2z = 0$ (d) $z = 4 - 2x - 3y$
    (e) $z = 4 + 2x + 3y$ (f) $x + y + z = 1$

    ANSWER:

    (a) and (b) are perpendicular, (a) and (f) are perpendicular, (c) and (e) are parallel.

    COMMENT:

    Follow-up Question. Does this mean that (b) is perpendicular to (f)? Why not?

3. Draw a picture and describe a situation where

    (a) The projection of a vector onto a line is the zero vector;
    (b) The projection of a vector onto a line is the vector itself.

    ANSWER:

    (a) The vector is perpendicular to the line.
    (b) The vector lies on the line.

    COMMENT:

    It is important to be able to visualize projections, not just remember the formula.

4. True or false?

    (a) $\vec{v} \cdot \vec{w} = 0$ if and only if $\vec{v} = \vec{0}$ or $\vec{w} = \vec{0}$.
    (b) The zero vector is orthogonal to any other vector.
    (c) Any plane has only two distinct normal vectors.
    (d) If $\|\vec{v}\|$ and $\|\vec{w}\|$ are large then $\vec{v} \cdot \vec{w}$ is large.
    (e) The dot product of a vector with itself is its magnitude.
    (f) Any vector normal to a surface has length one.
    (g) Parallel planes share a same normal vector.
    (h) If two planes are perpendicular, so are their normal vectors.

    ANSWER:

    (a) False, the vectors could be perpendicular.
    (b) True
    (c) False. A plane has infinitely many normal vectors. It has two distinct unit normal vectors.
    (d) False. Not if the vectors are perpendicular or close to perpendicular.
    (e) False. It's the square of its magnitude.
    (f) False
    (g) True
    (h) True

    COMMENT:

    This problem may bring up students' misconceptions about vectors.

ConcepTests and Answers and Comments for Section 13.4

1. Find $\vec{i} \times \vec{j}$ graphically. Don't use any formulas.

   ANSWER:

   By the right-hand-rule $\vec{i} \times \vec{j}$ points in the direction of $\vec{k}$. Since $\vec{i}$ and $\vec{j}$ both have length one and are perpendicular, they form the sides of a square of area one. Thus $||\vec{i} \times \vec{j}|| = 1$, so $\vec{i} \times \vec{j} = \vec{k}$.

   COMMENT:

   Follow-up Question. What about $\vec{j} \times \vec{i}$?

2. View the second and hour hands of a clock as vectors $\vec{s}$ and $\vec{h}$ in a horizontal plane in 3-space. Suppose $\vec{h}$ points at 12. At what time(s) during the next minute does $\vec{s} \times \vec{h}$ have largest magnitude?

   ANSWER:

   When the hands point in perpendicular directions, roughly after 15 and 45 seconds, the parallelogram that is formed by the two hands is largest.

   COMMENT:

   Follow-up Question. What is the difference in the cross product at the two times?

3. Decide if each of the following quantities is a vector, a scalar, or undefined.

   (a) $\vec{v} \cdot \vec{w}$
   (b) $\vec{v} \times \vec{w}$
   (c) $(\vec{u} \times \vec{v}) \cdot \vec{w}$
   (d) $\vec{u} \times (\vec{v} \cdot \vec{w})$
   (e) $\dfrac{\vec{v} \times \vec{w}}{\vec{v} \cdot \vec{w}}$
   (f) $\dfrac{\vec{v} \cdot \vec{w}}{\vec{v} \times \vec{w}}$

   ANSWER:

   (a) Scalar
   (b) Vector
   (c) Scalar
   (d) Undefined
   (e) Vector or undefined
   (f) Undefined

   COMMENT:

   $\dfrac{\vec{v} \times \vec{w}}{\vec{v} \cdot \vec{w}}$ is undefined when $\vec{v}$ and $\vec{w}$ are perpendicular.

4. Decide if each of the following statements is true or false.

   (a) If $||\vec{v}|| > ||\vec{w}||$ then $||\vec{v} \times \vec{u}|| > ||\vec{w} \times \vec{u}||$.
   (b) $\vec{u} \times \vec{u} = \vec{0}$ for all $\vec{u}$.
   (c) $\vec{u} \times \vec{v} = \vec{v} \times \vec{u}$
   (d) $||\vec{u} \times \vec{v}|| = ||\vec{v} \times \vec{u}||$
   (e) $(\vec{u} \times \vec{v}) \cdot \vec{u} = 0$

   ANSWER:

   (a) False
   (b) True
   (c) False
   (d) True
   (e) True

   COMMENT:

   Find a counterexample for each false statement. This encourages students to visualize.

Chapter Fourteen

ConcepTests and Answers and Comments for Section 14.1

Problems 1–2 concern the contour diagram for a function $f(x, y)$ in Figure 14.1.

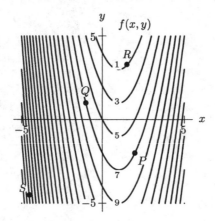

Figure 14.1

1. At the point Q, which of the following is true?

   (a) $f_x > 0, f_y > 0$ (b) $f_x > 0, f_y < 0$
   (c) $f_x < 0, f_y > 0$ (d) $f_x < 0, f_y < 0$

   ANSWER:
   (d), since both are negative.

2. List the points P, Q, R in order of increasing f_x.

   (a) $P > Q > R$ (b) $P > R > Q$ (c) $R > P > Q$
   (d) $R > Q > P$ (e) $Q > R > P$

   ANSWER:
   (b), since $f_x(P) > 0$, $f_x(R) > 0$, and $f_x(P) > f_x(R)$. Also $f_x(Q) < 0$.

3. Use the level curves of $f(x, y)$ in Figure 14.2 to estimate

(a) $f_x(2, 1)$ (b) $f_y(1, 2)$

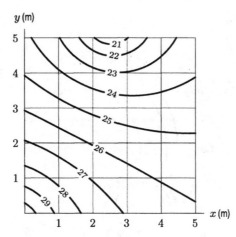

Figure 14.2

ANSWER:

(a) We have

$$f_x(2, 1) \approx \frac{26 - 27}{3.7 - 2} = -\frac{1}{1.7} = -0.588.$$

(b) We have

$$f_y(1, 2) \approx \frac{26 - 26.5}{2.5 - 2} = -1.$$

4. Figure 14.3 shows level curves of $f(x, y)$. At which of the following points is one or both of the partial derivatives, f_x, f_y, approximately zero? Which one(s)?

   (a) $(1, -0.5)$
   (b) $(-0.4, 1.5)$
   (c) $(1.5, -0.4)$
   (d) $(-0.5, 1)$

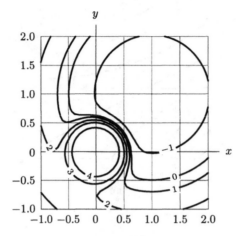

Figure 14.3

   ANSWER:

   (a) At $(1, -0.5)$, the level curve is approximately horizontal, so $f_x \approx 0$, but $f_y \not\approx 0$.
   (b) At $(-0.4, 1.5)$, neither are approximately zero.
   (c) At $(1.5, -0.4)$, neither are approximately zero.
   (d) At $(-0.5, 1)$, the level curve is approximately vertical, so $f_y \approx 0$, but $f_x \not\approx 0$.

   COMMENT:
   Ask students to estimate the nonzero partial derivative at one or more of the points.
   Ask students to speculate about the partial derivatives at $(0, 0)$.

5. Figure 14.4 is a contour diagram for $f(x, y)$ with the x and y axes in the usual directions. At the point P, is $f_x(P)$ positive or negative? Increasing or decreasing as we move in the the direction of increasing x?

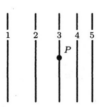

Figure 14.4

   ANSWER:
   At P, the values of f are increasing at an increasing rate as we move in the positive x-direction, so $f_x(P) > 0$, and $f_x(P)$ is increasing.

   COMMENT:
   Ask about how f_x changes in the y-direction and about f_y.

6. Figure 14.5 is a contour diagram for $f(x, y)$ with the x and y axes in the usual directions. At the point P, if x increases, what is true of $f_x(P)$? If y increases, what is true of $f_y(P)$?

(a) Have the same sign and both increase.
(b) Have the same sign and both decrease.
(c) Have opposite signs and both increase.
(d) Have opposite signs and both decrease.
(e) None of the above.

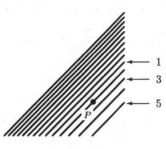

Figure 14.5

ANSWER:
We have $f_x(P) > 0$ and decreasing, and $f_y(P) < 0$ and decreasing, so the answer is (d).

ConcepTests and Answers and Comments for Section 14.2

1. Let $f(x, y) = x^2 \ln(x^2 y)$. Which of the following are the two partial derivatives of f? Which is which?

(a) $2x \ln(x^2 y) + 2x, \dfrac{2x}{y}$

(b) $\dfrac{1}{y}, 2x \ln(x^2 y) + \dfrac{2x}{y}$

(c) $\dfrac{x^2}{y}, 2x + 2x \ln(x^2 y)$

(d) $\dfrac{2x^3}{y} + 2x \ln(x^2 y), \dfrac{x^2}{y}$

(e) $\dfrac{2x}{y} + 2x \ln(x^2 y), \dfrac{1}{y}$

(f) $\dfrac{2x}{y}, 2x \ln(x^2 y) + \dfrac{2x^3}{y}$

ANSWER:
The partial derivatives are

$$f_x(x, y) = 2x \ln(x^2 y) + x^2 \cdot \frac{1}{x^2 y} \cdot 2xy = 2x \ln(x^2 y) + 2x$$

$$f_y(x, y) = x^2 \cdot \frac{1}{x^2 y} \cdot x^2 = \frac{x^2}{y}.$$

Thus, the answer is (c), with f_y first, f_x second.

2. Which of the following functions satisfy the following equation (called Euler's Equation):

$$xf_x + yf_y = f?$$

(a) x^2y^3 (b) $x + y + 1$ (c) $x^2 + y^2$ (d) $x^{0.4}y^{0.6}$

ANSWER:

Calculate the partial derivatives and check. The answer is (d), and

$$x\frac{\partial}{\partial x}(x^{0.4}y^{0.6}) = x0.4x^{-0.6}y^{0.6} = 0.4x^{0.4}y^{0.6}$$

$$y\frac{\partial}{\partial y}(x^{0.4}y^{0.6}) = yx^{0.4}0.6y^{-0.4} = 0.6x^{0.4}y^{0.6}.$$

Thus

$$xf_x + yf_y = 0.4x^{0.4}y^{0.6} + 0.6x^{0.4}y^{0.6} = x^{0.4}y^{0.6} = f.$$

ConcepTests and Answers and Comments for Section 14.3

1. Let $f(2,3) = 7$, $f_x(2,3) = -1$, and $f_y(2,3) = 4$. Then the tangent plane to the surface $z = f(x,y)$ at the point $(2,3)$ is

(a) $z = 7 - x + 4y$ (b) $x - 4y + z + 3 = 0$ (c) $-x + 4y + z = 7$
(d) $-x + 4y + z + 3 = 0$ (e) $z = 17 + x - 4y$

ANSWER:

The tangent plane is

$$z = 7 - 1(x - 2) + 4(y - 3),$$

which simplifies to

$$z = -3 - x + 4y \qquad \text{or} \qquad x - 4y + z + 3 = 0.$$

Thus, the answer is (b).

2. Figure 14.6 shows level curves of the function $f(x,y)$. The tangent plane approximation to $f(x,y)$ at the point P is

$$f(x,y) \approx c + m(x - x_0) + n(y - y_0).$$

What are the signs of c, m, and n?

(a) $c > 0, m > 0, n > 0$ (b) $c < 0, m > 0, n < 0$ (c) $c > 0, m < 0, n > 0$
(d) $c < 0, m < 0, n < 0$ (e) None of the above

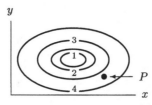

Figure 14.6

ANSWER:

Since $c = f(P) \approx 3.5$, we have $c > 0$. Since f increases in the x-direction and decreases in the y-direction, we know $m > 0$ and $n < 0$, so the answer is (e).

3. The differential of a function $f(x, y)$ at the point (a, b) is given by the formula

$$df = f_x(a, b)\, dx + f_y(a, b)\, dy.$$

Which of the following are true? (There may be more than one answer.)

(a) Doubling dx and dy doubles df.

(b) Moving to a different point (a, b) may change the formula for df.

(c) If dx and dy represent small changes in x and y in moving away from the point (a, b), then df approximates the change in f.

(d) The equation of the tangent plane to $z = f(x, y)$ at the point (a, b) can be used to calculate values of df from dx and dy.

(e) None of these are true.

ANSWER:

All of (a)–(d) are true.

4. A small business has \$300,000 worth of equipment and 100 workers. The total monthly production, P (in thousands of dollars), is a function of the total value of the equipment, V (in thousands of dollars), and the total number of workers, N. The differential of P is given by $dP = 4.9dN + 0.5dV$. If the business decides to lay off 3 workers and buy additional equipment worth \$20,000, then

(a) Monthly production increases.   (b) Monthly production decreases.

(c) Monthly production stays the same.

ANSWER:

(b), since $dN = -3$ and $dV = 20$, $dP = 4.9(-3) + 0.5(20) = -4.7$. Thus, we see that production, P, decreases (because dP is negative) by \$4700.

ConcepTests and Answers and Comments for Section 14.4

1. Figure 14.7 shows the temperature $T°C$ in a heated room as a function of distance x in meters along a wall and time t in minutes. Which is larger, $\| \operatorname{grad} T(15, 15)\|$ or $\| \operatorname{grad} T(25, 25)\|$? Explain your choice without computing these numbers.

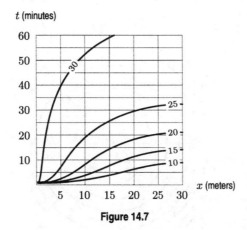

Figure 14.7

ANSWER:

We have $\| \operatorname{grad} T(15, 15)\| > \| \operatorname{grad} T(25, 25)\|$ because the contour lines are closer together at the point $(15, 15)$ than at $(25, 25)$.

2. Table 14.1 gives values of the function $f(x, y)$ which is smoothly varying around the point $(3, 5)$. Estimate the vector $\text{grad}(f(3, 5))$. If the gradient vector is placed with its tail at the origin, into which quadrant does the vector point?

(a) I   (b) II   (c) III   (d) IV

(e) Can't tell without more information

Table 14.1

		y	
		4.9	5	5.1
	2.9	18.12	17.42	16.73
x	3	18.42	17.74	17.04
	3.1	18.71	18.04	17.35

ANSWER:

Estimating the gradient from the table, we find

$$\frac{\partial f}{\partial x}(3, 5) \approx \frac{18.04 - 17.74}{3.1 - 3} = 3$$

$$\frac{\partial f}{\partial y}(3, 5) \approx \frac{17.04 - 17.74}{5.1 - 5} = -7.$$

Thus $\text{grad}(f(3, 5)) \approx 3\vec{i} - 7\vec{j}$, so $\text{grad}(f(3, 5))$ points into Quadrant IV, so the answer is (d).

3. Let $\text{grad} f(1, 1) = 3\vec{i} - 5\vec{j}$. What is the sign of the directional derivative of f in the directions given by each of the following vectors?

(a) ↖   (b) ↑   (c) ←   (d) ↘

ANSWER:

(a) The vector is approximately in the direction of $-\vec{i} + \vec{j}$, so the directional derivative is negative.

(b) The vector is approximately in the direction of $\vec{j}$, so the directional derivative is negative.

(c) The vector is approximately in the direction of $-\vec{i}$, so the directional derivative is negative.

(d) The vector is approximately in the direction of $\vec{i} - \vec{j}$, so the directional derivative is positive.

4. Which of the vectors shown on the contour diagram of $f(x, y)$ in Figure 14.8 could be $\text{grad} f$ at the point at which the tail is attached?

(a) A (b) B (c) C (d) D

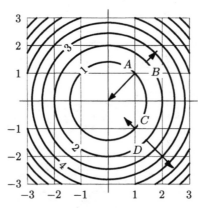

Figure 14.8

ANSWER:

(d). The vectors at A and C point in the opposite direction of $\text{grad} f$. The vector at B is too short. The vector at D could be $\text{grad} f$ at this point.

The magnitude of $\text{grad} f$ is the rate of change of f as we move perpendicular to the level curves. At B and D, we have $\| \text{grad} f \| \approx 1/0.5 = 2$.

5. At which of the points P, Q, R, S in Figure 14.9 does the gradient have the largest magnitude?

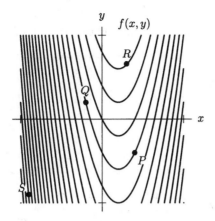

Figure 14.9

ANSWER:
At S, the level curves are closest together, so grad f has the largest magnitude.
COMMENT:
As an extension, ask students to label contours and show direction of grad f at each point.

6. The surface of a hill is modeled by $z = 25 - 2x^2 - 4y^2$. When a hiker reaches the point $(1, 1, 19)$, it begins to rain. She decides to descend the hill by the most rapid way. Which of the following vectors points in the direction in which she starts her descent?

(a) $-4x\vec{i} - 8y\vec{j}$ (b) $4x\vec{i} + 8y\vec{j}$ (c) $-4\vec{i} - 8\vec{j}$ (d) $4\vec{i} + 8\vec{j}$

(e) None of the above

ANSWER:
(d), since grad $z = -4x\vec{i} - 8y\vec{j}$, at the point $(1, 1, 19)$, we have grad $z = -4\vec{i} - 8\vec{j}$, which is the direction of most rapid ascent. To descend, she must go the opposite direction, in the direction of $4\vec{i} + 8\vec{j}$.

ConcepTests and Answers and Comments for Section 14.5

1. Let $f(x, y, z)$ represent the temperature in °C at the point (x, y, z) with x, y, z in meters. Let $\vec{v}$ be your velocity in meters per second. Give units and an interpretation of each of the following quantities.

(a) $\| \operatorname{grad} f \|$ (b) $\operatorname{grad} f \cdot \vec{v}$ (c) $\| \operatorname{grad} f \| \cdot \| \vec{v} \|$

ANSWER:

(a) The units of $\| \operatorname{grad} f \|$ are °C per meter. It represents the rate of change of temperature with distance as you move in the direction of grad f.

(b) The units of grad $f \cdot \vec{v}$ are °C per second. It represents the rate of change of temperature with time as you move with velocity $\vec{v}$.

(c) The units of $\| \operatorname{grad} f \| \cdot \| \vec{v} \|$ are °C per second. It represents the rate of change of temperature with time if you move in the direction of grad f with speed $\| \vec{v} \|$.

2. For $f(x, y, z)$, suppose

$$\text{grad } f(a, b, c) \cdot \vec{i} > \text{grad } f(a, b, c) \cdot \vec{j} > \text{grad } f(a, b, c) \cdot \vec{k} > 0.$$

The tangent plane to the surface $f(x, y, z) = 0$ through the point (a, b, c) is given by

$$z = p + mx + ny.$$

Which of the following is correct?

(a) $m > n > 0$ (b) $n > m > 0$ (c) $m < n < 0$ (d) $n < m < 0$

(e) None of the above

ANSWER:

The inequalities given tell us that

$$f_x(a, b, c) > f_y(a, b, c) > f_z(a, b, c) > 0.$$

The tangent plane to $f(x, y, z) = 0$ is given, for some constant d, by

$$f_x(a, b, c)x + f_y(a, b, c)y + f_z(a, b, c)z = d,$$

so

$$z = \frac{d}{f_z(a, b, c)} - \frac{f_x(a, b, c)}{f_z(a, b, c)}x - \frac{f_y(a, b, c)}{f_z(a, b, c)}y.$$

Thus,

$$m = -\frac{f_x(a, b, c)}{f_z(a, b, c)} \quad \text{and} \quad n = -\frac{f_y(a, b, c)}{f_z(a, b, c)},$$

so $m < n < 0$, and the answer is (c).

3. Let $\vec{r} = x\vec{i} + y\vec{j} + z\vec{k}$ and $\vec{r}_0 = x_0\vec{i} + y_0\vec{j} + z_0\vec{k}$. Which of the following is/are not a possible equation(s) for the tangent plane to the surface $f(x, y, z) = c$ at (x_0, y_0, z_0)? Why not?

(a) $\text{grad } f(x_0, y_0, z_0) \cdot \vec{r} = \text{grad } f(x_0, y_0, z_0) \cdot \vec{r}_0$

(b) $\text{grad } f(x_0, y_0, z_0) \times \vec{r} = \text{grad } f(x_0, y_0, z_0) \times \vec{r}_0$

(c) $z - f(x_0, y_0, z_0) = f_x(x_0, y_0, z_0)(x - x_0) + f_y(x_0, y_0, z_0)(y - y_0)$

(d) $f_x(x_0, y_0, z_0)x + f_y(x_0, y_0, z_0)y + f_z(x_0, y_0, z_0)z = f_x(x_0, y_0, z_0)x_0 + f_y(x_0, y_0, z_0)y_0 + f_z(x_0, y_0, z_0)z_0$

ANSWER:

Only (b) and (c) are incorrect. The equation for the tangent plane is a scalar equation; (b) is a vector equation. In (c), the left side should be $-f_z(x_0, y_0, z_0)(z - z_0)$.

COMMENT:

Have the students justify and explain the other equations.

4. The function $f(x, y)$ has gradient $\text{grad } f$ at the point (a, b). Which of the following statements is/are true?

(a) The vector $\text{grad } f$ is perpendicular to the level curve $f(x, y) = f(a, b)$.

(b) The vector $\text{grad } f$ is perpendicular to the surface $z = f(x, y)$ at the point $(a, b, f(a, b))$.

(c) The vector $f_x(a, b)\vec{i} + f_y(a, b)\vec{j} + \vec{k}$ is perpendicular to the surface $z = f(x, y)$.

(d) If the vector $\vec{n}$ is any vector which is perpendicular to the surface at the point where $x = a$ and $y = b$, then $\vec{n}$ is a scalar multiple of $(\text{grad } f - \vec{k})$.

ANSWER:

(a) True

(b) False. The vector $\text{grad } f$ is a 2-vector; the vector perpendicular to the surface has a z-component.

(c) False. The normal to the surface $z = f(x, y)$ is obtained by writing it in the form

$$f(x, y) - z = 0,$$

giving the normal as $f_x\vec{i} + f_y\vec{j} - \vec{k}$.

(d) True. One normal is $f_x\vec{i} + f_y\vec{j} - \vec{k} = \text{grad } f - \vec{k}$, so any other normal is a multiple of this one.

ConcepTests and Answers and Comments for Section 14.6

1. Figure 14.10 shows contours of $z = f(x, y)$. Figures 14.11 and 14.12 show x and y, respectively, as functions of t. Decide if $\left.\dfrac{dz}{dt}\right|_{t=2}$ is

   (a) Positive
   (b) Negative
   (c) Approximately 0
   (d) Can't tell without further information

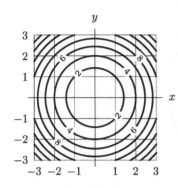

Figure 14.10

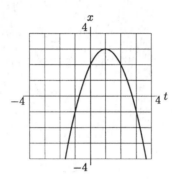

Figure 14.11

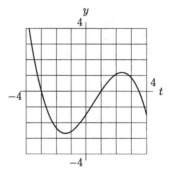

Figure 14.12

   ANSWER:
   By the chain rule

   $$\frac{dz}{dt} = z_x x'(t) + z_y y'(t).$$

   The graphs of $x(t)$ and $y(t)$ show that at $t = 2$, we have $x \approx 2$ and $y \approx 1$. In addition, $x'(2)$ is negative, and $y'(2)$ is positive, with $x'(2)$ having a much larger magnitude. We look at the point $(2, 1)$ on the contour diagram and see that $z_x(2, 1)$ and $z_y(2, 1)$ are both positive, with $z_x(2, 1)$ larger. Thus

   $$\left.\frac{dz}{dt}\right|_{t=2} = (+)(-) + (+)(+),$$

   and the relative magnitudes tells us that the first term, corresponding to $z_x x'(t)$, dominates, so $\left.\dfrac{dz}{dt}\right|_{t=2}$ is negative, and the answer is (b).

2. Let $z = g(u, v)$ and $u = u(x, y, t), v = v(x, y, t)$ and $x = x(t), y = y(t)$. Then the expression for dz/dt has

   (a) Three terms
   (b) Four terms
   (c) Six terms
   (d) Seven terms
   (e) Nine terms
   (f) None of the above

   ANSWER:
   By the chain rule,

   $$\frac{dz}{dt} = g_u u_x x' + g_u u_y y' + g_u u_t + g_v v_x x' + g_v v_y y' + g_v v_t.$$

   Thus, the answer is (c).

3. Let $s = f(x, y, z)$ and $x = x(u, v, w), y = y(u, v, w), z = z(u, v, w)$. To calculate $\dfrac{\partial s}{\partial u}(1, 2, 3)$, which of the following pieces of data do you **not** need?

   (a) $f(1, 2, 3) = 5$
   (b) $f(7, 8, 9) = 6$
   (c) $x(1, 2, 3) = 7$
   (d) $y(1, 2, 3) = 8$
   (e) $z(1, 2, 3) = 9$
   (f) $f_x(1, 2, 3) = 20$
   (g) $f_x(7, 8, 9) = 30$
   (h) $x_u(1, 2, 3) = -5$
   (i) $x_u(7, 8, 9) = -7$

   ANSWER:
   Since we evaluate $\partial s / \partial u$ at $(1, 2, 3)$, we are working at $u = 1, v = 2, w = 3$. We need the corresponding values of x, y, z which are $x = 7, y = 8, z = 9$. In the formula for the chain rule, the partial derivatives of f are evaluated at $(7, 8, 9)$, and the partial derivatives of x, y, z are evaluated at $(1, 2, 3)$. Thus, we need (c), (d), (e), (g), and (h), but not (a), (b), (f), (i).

ConcepTests and Answers and Comments for Section 14.7

1. Figure 14.13 shows level curves of $f(x, y)$. What are the signs of $f_{xx}(P)$ and $f_{yy}(P)$?

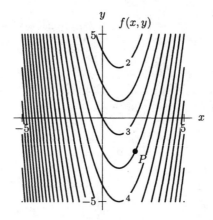

Figure 14.13

ANSWER:
Since f_x is positive and increasing at P, we have $f_{xx}(P) > 0$. Since f_y is negative, but getting less negative at P, we have $f_{yy}(P) > 0$. However, f_y is changing very slowly, so it would also be reasonable to say $f_{yy}(P) \approx 0$.
COMMENT:
As an extension, ask students to find the sign of $f_{xy}(P)$.

2. Figure 14.14 shows level curves of $f(x, y)$. What are the signs of $f_{xx}(Q)$ and $f_{yy}(Q)$?

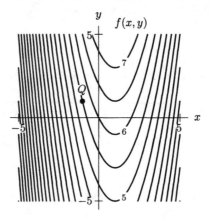

Figure 14.14

ANSWER:
Since f_x is positive and decreasing at Q, we have $f_{xx}(Q) < 0$. Since f_y is positive and decreasing at Q, we have $f_{yy}(Q) < 0$. However, f_y is changing very slowly, so it would also be reasonable to say $f_{yy}(Q) \approx 0$.
COMMENT:
As an extension, ask students to find the sign of $f_{xy}(Q)$.

3. Figure 14.15 shows the surface $z = f(x, y)$. What are the signs of $f_{xx}(A)$ and $f_{yy}(A)$?

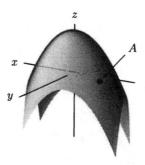

Figure 14.15

ANSWER:

Since $f_x(A)$ is positive and decreasing as we move in the direction of increasing x, we have $f_{xx}(A) < 0$. Since $f_y(A)$ is negative and decreasing as we move in the direction of increasing y, we have $f_{yy}(A) < 0$.

COMMENT:

Ask students to think about the sign of $f_{xy}(A)$.

4. Figure 14.16 shows the surface $z = f(x, y)$. What are the signs of $f_{xx}(P)$ and $f_{yy}(P)$?

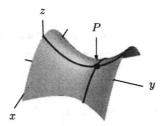

Figure 14.16

ANSWER:

Since $f_x(P)$ is decreasing in the x-direction, $f_{xx}(P) < 0$. Since $f_y(P)$ is increasing in the y-direction, $f_{yy}(P) > 0$.

5. Figure 14.17 shows the temperature $T°C$ as a function of distance x in meters along a wall and time t in minutes. Choose the correct statement and explain your choice without computing these partial derivatives.

(a) $\dfrac{\partial T}{\partial t}(t, 10) < 0$ and $\dfrac{\partial^2 T}{\partial t^2}(t, 10) < 0.$ (b) $\dfrac{\partial T}{\partial t}(t, 10) > 0$ and $\dfrac{\partial^2 T}{\partial t^2}(t, 10) > 0.$

(c) $\dfrac{\partial T}{\partial t}(t, 10) > 0$ and $\dfrac{\partial^2 T}{\partial t^2}(t, 10) < 0.$ (d) $\dfrac{\partial T}{\partial t}(t, 10) < 0$ and $\dfrac{\partial^2 T}{\partial t^2}(t, 10) > 0.$

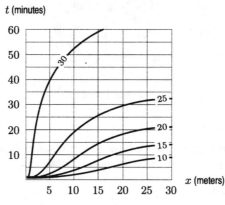

Figure 14.17

ANSWER:

(c). From the graph of the contour lines, we can see that along the line $x = 10$, the temperature T is increasing, thus $\dfrac{\partial T}{\partial t}(t, 10) > 0$. Since the distance between the contour lines is increasing as time increases along the line $x = 10$, the rate of change of $\dfrac{\partial T}{\partial t}(t, 10)$ is decreasing, so $\dfrac{\partial^2 T}{\partial t^2}(t, 10) < 0.$

6. The quadratic Taylor Polynomials (a)–(d) each approximate a function of two variables near the origin. Figures (I)–(IV) are contours near the origin. Match (a)–(d) to (I)–(IV).

(a) $-x^2 + y^2$ (b) $x^2 - y^2$ (c) $-x^2 - y^2$ (d) $x^2 + y^2$

(I)

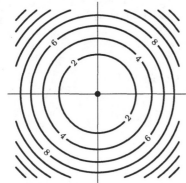

(II)

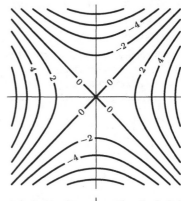

(III)

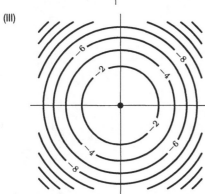

(IV)

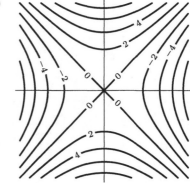

ANSWER:

(a) corresponds to (IV).
(b) corresponds to (II).
(c) corresponds to (III).
(d) corresponds to (I).

ConcepTests and Answers and Comments for Section 14.8

1. Which of the following functions $f(x, y)$ is differentiable at the given point?

   (a) $\sqrt{1 - x^2 - y^2}$ at $(0, 0)$

   (b) $\sqrt{4 - x^2 - y^2}$ at $(2, 0)$

   (c) $-\sqrt{x^2 + 2y^2}$ at $(0, 0)$

   (d) $-\sqrt{x^2 + 2y^2}$ at $(2, 0)$

   ANSWER:

   (a) Differentiable; point is at top of hemisphere.
   (b) Not differentiable; hemisphere vertical at this point.
   (c) Not differentiable; top point of cone with ellipse-shaped cross-section.
   (d) Differentiable; point on side of cone.

Chapter Fifteen

ConcepTests and Answers and Comments for Section 15.1 ――――――――

Problems 1–3 refer to the functions $f(x, y)$:

(a) $x^2 + 2y^3$

(b) $x^2y + 4xy + 4y$

(c) $x^2y^3 - x^4 + 2y$

(d) $x \cos y$

1. Which of these functions has a critical point at the origin?

    ANSWER:

    (a). Taking partial derivatives gives $f_x = 2x$ and $f_y = 6y^2$, so $x = 0$ and $y = 0$ give a critical point. The other functions do not have a critical point at the origin.

2. True or False? Function (b) has a local maximum at the origin.

    ANSWER:

    False. Since $f_y(0, 0) = 4$, the origin is not a critical point of f, so it cannot be a local maximum.

3. Which of the functions does not have a critical point?

    ANSWER:

    (c). Taking partial derivatives gives $f_x = 2xy^3 - 4x^3$ and $f_y = 3x^2y^2 + 2$. Because f_y is always greater than or equal to 2, there are no critical points.

Problems 4–7 refer to the functions $f(x, y)$:

(a) $x^2 + 2x + 2y^3 - y^2$

(b) $x^2y + xy$

(c) $x^2y^2 - (1/2)x^4 + 2y$

(d) $x^4y - 7y$

4. Which of the functions has a critical point at the origin?

    ANSWER:

    (b). Taking partial derivatives gives $f_x = 2xy + y = (2x + 1)y$ and $f_y = x^2 + x = (x + 1)x$, so $x = 0$ and $y = 0$ give a critical point. The others do not have a critical point at the origin.

5. True or false? The value of the discriminant for function (b) determines that the origin is a local maximum.

    ANSWER:

    False. Taking partial derivatives gives $f_{xx} = 2y$, $f_{yy} = 0$, and $f_{xy} = 2x + 1$, so at $(0, 0)$ the discriminant

    $$D = (2 \cdot 0) \cdot 0 - (2 \cdot 0 + 1)^2 = -1,$$

    giving a saddle point at $(0, 0)$.

6. Which of these functions does not have a critical point with $y = 0$?

    ANSWER:

    (c). Taking partial derivatives gives $f_x = 2xy^2 - 2x^3 = 2x(y^2 - x^2)$ and $f_y = 2x^2y + 2 = 2(x^2y + 1)$. At critical points, we need $y^2 = x^2$ and $x^2y = -1$, so $y^3 = -1$, or $y = -1$. Thus, x may equal either 1 or -1, so the only two critical points are $(1, -1)$ and $(-1, -1)$.

7. Which of these functions does not have a critical point with $x = -1$?

    ANSWER:

    (d). Taking partial derivatives gives $f_x = 4x^3y$ and $f_y = x^4 - 7$, so the two critical points are $x = \pm 7^{1/4}, y = 0$.

ConcepTests and Answers and Comments for Section 15.2

1. Estimate the global maximum and minimum of the functions whose level curves are in Figure 15.1. How many times does each occur?

   (a) Max = 6, occurring once; min = −6, occurring once
   (b) Max = 6, occurring once; min = −6, occurring twice
   (c) Max = 6, occurring twice; min = −6, occurring twice
   (d) Max = 6, occurring three times; min = −6, occurring three times
   (e) None of the above

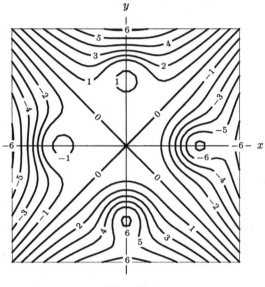

Figure 15.1

ANSWER:

The global max is about 6, or slightly higher. It occurs three times, twice on the negative y-axis and once on the positive y-axis. The global min is −6, or slightly lower. It occurs three times, twice on the positive x-axis and once on the negative x-axis. The answer is (d).

2. What are the global maximum and minimum values of $f(x, y) = x^2 + y^2$ on the triangular region in the first quadrant bounded by $x + y = 2$, $x = 0$, $y = 0$?

   ANSWER:

   The minimum value of $x^2 + y^2$ occurs at the origin, so the minimum is 0. The maximum occurs at the other two corners of the triangle, so at $(2, 0)$ and $(0, 2)$; the maximum value is 4.

   COMMENT:

   As an extension, ask for the global maximum and minimum on the hypotenuse of the triangle. Then, the maximum remains the same, but the minimum becomes 2. This question can be used as a lead-in to Lagrange multipliers.

ConcepTests and Answers and Comments for Section 15.3 ──────────

For Problems 1–2, use Figure 15.2. The grid lines are one unit apart.

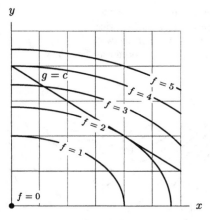

Figure 15.2

1. Find the maximum and minimum values of f on $g = c$. At which points do they occur?
   ANSWER:
   The maximum is $f = 4$ and occurs at $(0, 4)$. The minimum is $f = 2$ and occurs at about $(4, 2)$.

2. Find the maximum and minimum values of f on the triangular region below $g = c$ in the first quadrant.
   ANSWER:
   The maximum is $f = 4$ and occurs at $(0, 4)$. The minimum is $f = 0$ and occurs at the origin.

Problems 3–4 concern the following:

(a) Maximize x^2y^2 subject to $x^2 + y^2 = 4$.

(b) Maximize $x + y$ subject to $x^2 + y^2 = 4$.

(c) Minimize $x^2 + y^2$ subject to $x^2y^2 = 4$.

(d) Maximize x^2y^2 subject to $x + y = 4$.

3. Which one of (a)–(d) does not have its solution at the same point in the first quadrant as the others?

   ANSWER:

   The answer is (d).

   (a) We solve

   $$2xy^2 = 2x\lambda$$
   $$2x^2y = 2y\lambda$$
   $$x^2 + y^2 = 4,$$

   giving $x^2 = y^2$, so $2x^2 = 4$, so (since we are working in the first quadrant) $x = y = \sqrt{2}$.

   (b) We solve

   $$1 = 2x\lambda$$
   $$1 = 2y\lambda$$
   $$x^2 + y^2 = 4,$$

   giving $x = y$, so $2x^2 = 4$, so (since we are working in the first quadrant) $x = y = \sqrt{2}$.

   (c) We solve

   $$2x = 2xy^2\lambda$$
   $$2y = 2x^2y\lambda$$
   $$x^2y^2 = 4,$$

   giving $x^2 = y^2$, so $(x^2)^2 = 4$, so (since we are working in the first quadrant) $x = y = \sqrt{2}$.

   (d) We solve

   $$2xy^2 = \lambda$$
   $$2x^2y = \lambda$$
   $$x + y = 4.$$

   Dividing the first two equations gives $x = y$, so (since we are working in the first quadrant) $x = y = 2$.

   Thus, (a), (b), and (c) have their solution at $x = y = \sqrt{2}$, and (d) has its solution at a different point, namely $x = y = 2$.

4. Which two of (a)–(d) have the same extreme value of the objective function in the first quadrant?

   ANSWER:

   (a) We have $f(\sqrt{2}, \sqrt{2}) = (\sqrt{2})^2(\sqrt{2})^2 = 4$.
   (b) We have $f(\sqrt{2}, \sqrt{2}) = \sqrt{2} + \sqrt{2} = 2\sqrt{2}$.
   (c) We have $f(\sqrt{2}, \sqrt{2}) = (\sqrt{2})^2 + (\sqrt{2})^2 = 4$.
   (d) We have $f(2, 2) = 2^2 \cdot 2^2 = 16$.

   Thus, (a) and (c) have the same extreme value.

5. Find the maximum of the production function $f(x, y) = xy$ in the first quadrant subject to each of the three budget constraints.

    I. $x + y = 12$ II. $2x + 5y = 12$ III. $3x + y/2 = 12$

Arrange the x- and y-coordinates of the optimal point in increasing order. Pick one of (a)–(e) and one of (f)–(j).

(a) x: I < II < III

(b) x: III < II < I

(c) x: II < III < I

(d) x: II < I < III

(e) x: III < I < II

(f) y: I < II < III

(g) y: III < II < I

(h) y: II < III < I

(i) y: II < I < III

(j) y: III < I < II

ANSWER:

For (I), we have

$$y = \lambda$$
$$x = \lambda$$
$$x + y = 12,$$

so $x = y$ and the solution is $x = 6$, $y = 6$.

For (II), we have

$$y = 2\lambda$$
$$x = 5\lambda$$
$$2x + 5y = 12,$$

so $2x = 5y$, and the solution is $x = 3, y = 1.2$.

For (III), we have

$$y = 3\lambda$$
$$x = \lambda/2$$
$$3x + y/2 = 12,$$

so $3x = y/2$, and the solution is $x = 2, y = 12$.

Thus, for x: III<II<I

        for y: II<I<III

The answer is (b) and (i).

6. The point P is a maximum or minimum of the function f subject to the constraint $g(x, y) = x + y = c$, with $x, y \geq 0$. For each of the cases, does P give a maximum or a minimum of f? What is the sign of λ? If P gives a maximum, where does the minimum of f occur? If P gives a minimum, where does the maximum of f occur?

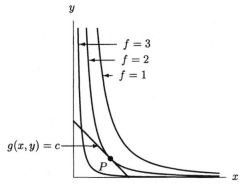

Figure 15.3

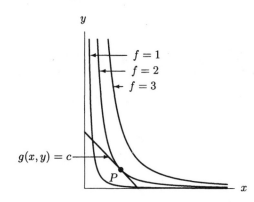

Figure 15.4

ANSWER:

(a) The point P gives a minimum; the maximum is at one of the end points of the line segment (either the x- or the y-intercept). The value of λ is negative, since f decreases in the direction in which g increases.

(b) The point P gives a maximum; the minimum is at the x- or y-intercept. The value of λ is positive, since f and g increase in the same direction.

7. Figure 15.5 shows the optimal point (marked with a dot) in three optimization problems with the same constraint. Arrange the corresponding values of λ in increasing order. (Assume λ is positive.)

(a) I<II<III   (b) II<III<I   (c) III<II<I   (d) I<III<II

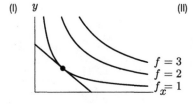   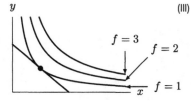   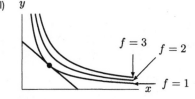

Figure 15.5

ANSWER:

Since λ is the additional f that is obtained by relaxing the constraint by 1 unit, λ is larger if the level curves of f are close together near the optimal point. The answer is I<II<III, so (a).

Chapter Sixteen

ConcepTests and Answers and Comments for Section 16.1

1. Let R be the region $0 \le x \le 20, 0 \le y \le 40$. The function $f(x, y)$ represents the performance index of a machine; its values range from 0 to 1. Use the grid in Figure 16.1, which is a contour map of f, to make upper and lower estimates of

$$I = \int_R f(x, y)\, dA.$$

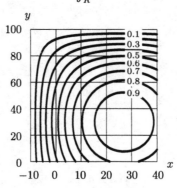

Figure 16.1

ANSWER:
The grid has $\Delta x = 10, \Delta y = 20$.

$$\text{Lower estimate} = (0.5 + 0.8 + 0.4 + 0.7)10 \cdot 20 = 480$$
$$\text{Upper estimate} = (0.8 + 0.8 + 1 + 1)10 \cdot 20 = 720$$

Many other answers are possible.

Problems 2–3 concern the integrals

(a) $\int_R x\, dA$ (b) $\int_R y\, dA$ (c) $\int_T y\, dA$ (d) $\int_R (x - x^2)\, dA$

(e) $\int_T (y - y^2) A$ (f) $\int_L (x^2 - x)\, dA$ (g) $\int_L (y + y^3)\, dA$ (h) $\int_R (2x + 3y)\, dA$

Here R is the rectangle $-1 \le x \le 1, -1 \le y \le 1$, and T is the top half $-1 \le x \le 1, 0 \le y \le 1$, and L the left half $-1 \le x \le 0, -1 \le y \le 1$.

2. Without evaluating them, decide which of the integrals are positive, which are negative, and which are zero.
    ANSWER:
    Positive: (c), (e), (f)
Negative: (d), (g)
Zero: (a), (b), (h)

3. Without evaluating them, arrange the integrals (a), (c), (e), (f) in ascending order.
    ANSWER:
    By symmetry, (a)$= \int_R x\, dA = 0$.
    Since the region of integration, T, is the same in (c) and (e), and since $0 \le y^2 \le y$ for $0 \le y \le 1$, we see that $0 \le y - y^2 \le y$ for points in T. Thus

$$\int_T (y - y^2)\, dA < \int_T y\, dA, \qquad \text{so (e)} < \text{(c)}.$$

For (f), the region of integration is L, where $-1 \le x \le 0$, so $x^2 - x \ge 0$. Since both x^2 and $-x$ are positive on L, we have

$$\int_L -x\, dA \le \int_L (x^2 - x)\, dA.$$

But $\int_L -x\, dA = \int_T y\, dA$, so (c)<(f). Putting these results together, we have

$$\text{(a)} < \text{(e)} < \text{(c)} < \text{(f)}.$$

4. Let R be the region $10 \leq x \leq 14, 20 \leq y \leq 30$. Table 16.1 gives values of $f(x, y)$. Using upper and lower Riemann sums, what are the best possible upper and lower estimates for the integral

$$I = \int_R f(x, y) \, dx \, dy.$$

(a) $23 < I < 990$
(b) $92 < I < 300$
(c) $160 < I < 396$
(d) $160 < I < 300$
(e) $92 < I < 396$

Table 16.1

		\multicolumn{3}{c}{y}		
		20	25	30
	10	2.3	4.2	7.3
x	12	3.7	5.8	8.1
	14	4.3	6.2	9.9

ANSWER:
(d). The function appears to be increasing in x and in y, so since $\Delta x \Delta y = 2 \cdot 5 = 10$

$$\text{Lower estimate} = (2.3 + 3.7 + 4.2 + 5.8)\Delta x \Delta y = 16 \cdot 10 = 160.$$
$$\text{Upper estimate} = (5.8 + 6.2 + 8.1 + 9.9)\Delta x \Delta y = 30 \cdot 10 = 300.$$

So $160 \leq I \leq 300$.

ConcepTests and Answers and Comments for Section 16.2

1. The integral $\int_0^1 \int_0^1 x^2 \, dx \, dy$ represents the

   (a) Area under the curve $y = x^2$ between $x = 0$ and $x = 1$.
   (b) Volume under the surface $z = x^2$ above the square $0 \leq x, y \leq 1$ on the xy-plane.
   (c) Area under the curve $y = x^2$ above the square $0 \leq x, y \leq 1$ on the xy-plane.

   ANSWER:
   (b). The volume under the surface $z = x^2$ above the square $0 \leq x, y \leq 1$ on the xy-plane.
   COMMENT:
   Students may think that any integral represents an area.

2. The integral $\int_0^1 \int_x^1 \, dy \, dx$ represents the

   (a) Area of a triangular region in the xy-plane.
   (b) Volume under the plane $z = 1$ above a triangular region of the plane.
   (c) Area of a square in the xy-plane.

   ANSWER:
   (a) and (b) are both right.
   COMMENT:
   Students may think that an integral can only have one interpretation.

3. Let $f(x, y)$ be a positive function. Rank the following integrals from smallest to largest.

   (a) $\int_0^1 \int_{x^2}^1 f(x, y) \, dy \, dx$,
   (b) $\int_0^1 \int_{x^3}^1 f(x, y) \, dy \, dx$,
   (c) $\int_0^1 \int_0^1 f(x, y) \, dy \, dx$.

   ANSWER:
   (a) is smaller than (b) which is smaller than (c). Since the regions are nested and the integrands are the same and positive, the only thing that matters is the size of the region we integrate over.

4. Match the integral with the appropriate region of integration.

   (a) R_1 : The triangle with vertices $(0,0)$, $(2,0)$, $(0,1)$.
   (b) R_2 : The triangle with vertices $(0,0)$, $(0,2)$, $(1,0)$.
   (c) R_3 : The triangle with vertices $(0,0)$, $(2,0)$, $(2,1)$.
   (d) R_4 : The triangle with vertices $(0,0)$, $(1,0)$, $(1,2)$.

   (i) $\int_0^1 \int_0^{2-2x} f(x,y) \, dy \, dx$
   (ii) $\int_0^1 \int_0^{2-2y} f(x,y) \, dx \, dy$
   (iii) $\int_0^1 \int_0^{2x} f(x,y) \, dy \, dx$
   (iv) $\int_0^1 \int_{2y}^{2} f(x,y) \, dx \, dy$

   ANSWER:
   (a) and (ii); (b) and (i); (c) and (iv); (d) and (iii)
   COMMENT:
   Follow-up Question. Write each integral with the order of integration reversed.

5. Which of the following integrals is (are) equal to $\int_0^3 \int_0^{4x} f(x,y) \, dy \, dx$?

   (a) $\int_0^{4x} \int_0^3 f(x,y) \, dx \, dy$
   (b) $\int_0^{12} \int_{y/4}^3 f(x,y) \, dx \, dy$
   (c) $\int_0^{12} \int_3^{y/4} f(x,y) \, dx \, dy$
   (d) $\int_0^{12} \int_0^{y/4} f(x,y) \, dx \, dy$
   (e) $\int_0^{4x} \int_0^3 f(x,y) \, dy \, dx$

   ANSWER:
   (b) $\int_0^{12} \int_{y/4}^3 f(x,y) \, dx \, dy$
   COMMENT:
   Follow-up Question. Which of these integrals don't make sense?

ConcepTests and Answers and Comments for Section 16.3

1. What does the integral $\int_0^1 \int_0^1 \int_0^1 z \, dz \, dy \, dx$ represent?

   (a) The volume of a cube of side 1.
   (b) The volume of a sphere of radius 1.
   (c) The area of a square of side 1.
   (d) None of the above.

   ANSWER:
   (d) None of the above.
   COMMENT:
   Follow-up Question. What could it represent?

2. Set up a single, double, and triple integral, each of which gives the volume of the region under the plane $z = x$ and above the rectangle $0 \le x \le 1, 0 \le y \le 2$. Explain what each integrand represents.

   ANSWER:
   $\int_0^1 2x \, dx$; here, $2x$ is the area of a rectangular cross section of the three dimensional region parallel to the yz-plane. Then, $2x \, dx$ is the volume of a slab parallel to the yz-plane, but only infinitesimally thick in the x-direction.

   $\int_0^1 \int_0^2 x \, dy \, dx$; here x is the length of a vertical line segment from the bottom to the top of the region. Then, $x \, dy \, dx$ is the volume of a vertical column that is infinitesimally wide in the x- and y-directions.

   $\int_0^1 \int_0^2 \int_0^x \, dz \, dy \, dx$, here the entire region is described by the limits of integration. Then, $dz \, dy \, dx$ represents the volume of an infinitesimal box anywhere in the region.

   COMMENT:
   If you start with the triple integral, the other integrals are stages of the iterative solution.

3. For each integral (a)–(d) that makes sense, match it with its region of integration, I or II.

   (a) $\int_1^3 \int_{y-1}^2 \int_0^y f(x,y,z)\, dz\, dx\, dy$

   (b) $\int_1^3 \int_0^y \int_2^{y-1} f(x,y,z)\, dx\, dy\, dz$

   (c) $\int_{-1}^1 \int_{-\sqrt{1-x^2}}^{\sqrt{1-x^2}} \int_0^{\sqrt{1-x^2-y^2}} f(x,y,z)\, dz\, dy\, dx$

   (d) $\int_{-1}^1 \int_{-\sqrt{1-y^2}}^{\sqrt{1-y^2}} \int_0^{\sqrt{1-x^2-y^2}} f(x,y,z)\, dz\, dx\, dy$

   I The region below the plane $z = y$ above the triangle with vertices $(0,1)$, $(2,1)$, $(2,3)$ in the xy-plane.

   II The region between the upper hemisphere of $x^2 + y^2 + z^2 = 1$ and the xy-plane.

   ANSWER:
   I is the region of integration of the first integral, (a).
   II is the region of integration for the last two integrals, (c) and (d).
   COMMENT:
   Follow-up Question. What is wrong with the second integral?

4. Which of the following integrals is (are) equal to $\int_0^3 \int_0^2 \int_0^y f(x,y,z)\, dz\, dy\, dx$?

   (a) $\int_0^2 \int_0^3 \int_0^y f(x,y,z)\, dz\, dx\, dy$

   (b) $\int_0^2 \int_0^3 \int_0^y f(x,y,z)\, dz\, dy\, dx$

   (c) $\int_0^3 \int_0^2 \int_0^y f(x,y,z)\, dx\, dy\, dz$

   (d) $\int_0^3 \int_0^2 \int_z^2 f(x,y,z)\, dy\, dz\, dx$

   (e) $\int_0^3 \int_0^2 \int_0^z f(x,y,z)\, dy\, dz\, dx$

   ANSWER:
   (a) $\int_0^2 \int_0^3 \int_0^y f(x,y,z)\, dz\, dx\, dy$ and (d) $\int_0^3 \int_0^2 \int_z^2 f(x,y,z)\, dy\, dz\, dx$
   COMMENT:
   Follow-up Question. What is the region of integration of the other integrals?

5. $\int_{-1}^1 \int_{-\sqrt{1-x^2}}^{\sqrt{1-x^2}} \int_{-\sqrt{1-x^2-y^2}}^{\sqrt{1-x^2-y^2}} (x^2 + y^2 + z^2)\, dz\, dy\, dx$ describes the mass of

   (a) A cone that gets heavier toward the outside.
   (b) A cone that gets lighter toward the outside.
   (c) A ball that gets heavier toward the outside.
   (d) A ball that gets lighter toward the outside.

   ANSWER:
   (c) A ball that gets heavier toward the outside.
   COMMENT:
   Follow-up Question. How would the integral have to change to represent each of the other choices?

ConcepTests and Answers and Comments for Section 16.4

1. What geometric shape (i)–(iv) is described by the equations in (a)–(f)?

   (i) line   (ii) circle   (iii) spiral   (iv) none of the above

   (a) $r = \theta$ (b) $r = 4$ (c) $\theta = 4$

   (d) $r = 1/\theta$ (e) $r = \sin\theta$ (f) $r = 1/\sin\theta$

   ANSWER:
   (a) is (iii), spiral
   (b) is (ii), circle
   (c) is (i), line
   (d) is (iii), spiral
   (e) is (ii), circle
   (f) is (i), line
   COMMENT:
   Follow-up Question. How do the two lines, the two spirals, and the two circles differ?

2. Describe each of the following regions in words:

   (a) $0 \le r \le 2, 0 \le \theta \le 2\pi$
   (b) $0 \le r < \infty, 0 \le \theta \le \pi$
   (c) $0 \le r \le \theta, 0 \le \theta \le \pi$
   (d) $2 \le r \le 4, \pi \le \theta \le 3\pi/2$

   ANSWER:

   (a) Disk of radius 2, centered at the origin.
   (b) Upper half of xy-plane.
   (c) Region in the upper half plane beneath the first half-turn of the spiral $r = \theta, \theta \ge 0$.
   (d) Sector of an annulus in the third quadrant.

   COMMENT:
   Follow-up Question. Set up integrals that compute the areas of each region.

3. Which of the following integrals give the area of the unit circle?

   (a) $\int_{-1}^{1} \int_{-\sqrt{1-x^2}}^{\sqrt{1-x^2}} dy\, dx$
   (b) $\int_{-1}^{1} \int_{-\sqrt{1-x^2}}^{\sqrt{1-x^2}} x\, dy\, dx$
   (c) $\int_0^{2\pi} \int_0^1 r\, dr\, d\theta$
   (d) $\int_0^{2\pi} \int_0^1 dr\, d\theta$
   (e) $\int_0^1 \int_0^{2\pi} r\, d\theta\, dr$
   (f) $\int_0^1 \int_0^{2\pi} d\theta\, dr$

   ANSWER:
   (a), (c), (e)

   COMMENT:
   Follow-up Question. Where does the r in the correct polar integrals come from?

4. Describe the region of integration for $\int_{\pi/4}^{\pi/2} \int_{1/\sin\theta}^{4/\sin\theta} f(r, \theta) r\, dr\, d\theta$.

   ANSWER:
   The region lies in the first quadrant and is bounded by four lines. The equations $r = 1/\sin\theta$ and $r = 4/\sin\theta$ are the horizontal lines $y = r\sin\theta = 1$ and $y = r\sin\theta = 4$. The equation $\theta = \pi/4$ gives the line $y = x$, and the equation $\theta = \pi/2$ gives the y-axis.

   COMMENT:
   Follow-up Question. Rewrite the integral in Cartesian coordinates.

ConcepTests and Answers and Comments for Section 16.5

1. Each of the following, (a)–(f), represents a point, a curve, a surface, or a solid region in cylindrical or spherical coordinates. Decide which it represents and describe the region in words.

   (a) $0 \le \rho < \infty, \theta = \pi, 0 \le \phi \le \pi$
   (b) $r = 3, 0 \le \theta \le 2\pi, -\infty < z < \infty$
   (c) $r = 3, \theta = \frac{\pi}{2}, -\infty < z < \infty$
   (d) $1 \le r \le 4, 0 \le \theta \le 2\pi, -5 \le z \le 2$
   (e) $1 \le \rho \le 4, 0 \le \theta \le 2\pi, 0 \le \phi \le \pi$
   (f) $\rho = 1, \theta = 3, \phi = 2$

   ANSWER:

   (a) Surface. This is the half-plane $y = 0, x \le 0$, which is vertical and perpendicular to the y-axis.
   (b) Surface of the cylinder of radius 3 centered on the z-axis.
   (c) Line, parallel to the z-axis, with $x = 0, y = 3$.
   (d) Solid region. A solid cylinder of radius 4, centered on z-axis from $z = -5$ to $z = 2$, with the central cylindrical core core of radius 1 removed.
   (e) Solid region. A solid ball of radius 4, centered at the origin, with the smaller ball of radius 1 removed from its interior.
   (f) Point $(-0.90, 0.13, -0.42)$, since $x = \cos 3 \sin 2 = -0.90, y = \sin 3 \sin 2 = 0.13, z = \cos 2 = -0.42$.

   COMMENT:
   Follow-up Question. Can you describe the objects in another coordinate system?

2. Describe the region of integration of $\int_0^2 \int_{-\pi/4}^{\pi/4} \int_0^3 f(r, \theta, z) r \, dr \, d\theta \, dz$.

ANSWER:

The region is a $90°$ wedge of a solid cylinder of radius 3 and height 2, centered around the z-axis, sitting on the plane $z = 0$. Laterally, the wedge is bounded by the vertical planes $\theta = -\pi/4$ and $\theta = \pi/4$. The edge of the wedge is on the z-axis.

COMMENT:

Follow-up Question. Why do you want to use cylindrical coordinates to describe this region?

3. Which of the following integrals give the volume of the unit sphere?

(a) $\int_0^{2\pi} \int_0^{2\pi} \int_0^1 d\rho \, d\theta \, d\phi$

(b) $\int_0^{\pi} \int_0^{2\pi} \int_0^1 d\rho \, d\theta \, d\phi$

(c) $\int_0^{\pi} \int_0^{2\pi} \int_0^1 \rho^2 \sin \phi \, d\rho \, d\theta \, d\phi$

(d) $\int_0^{\pi} \int_0^{2\pi} \int_0^1 \rho^2 \sin \phi \, d\rho \, d\phi \, d\theta$

(e) $\int_0^{\pi} \int_0^{2\pi} \int_0^1 \rho \, d\rho \, d\phi \, d\theta$

ANSWER:

(c) $\int_0^{\pi} \int_0^{2\pi} \int_0^1 \rho^2 \sin \phi \, d\rho \, d\theta \, d\phi$

COMMENT:

Follow-up Question. Why are spherical coordinates called spherical coordinates?

ConcepTests and Answers and Comments for Section 16.6

For Problems 1–7, let $p(x, y)$ be a joint density function for x and y. Are the following statements true or false?

1. $\int_a^b \int_{-\infty}^{\infty} p(x, y) \, dy \, dx$ is the probability that $a \leq x \leq b$.

ANSWER:

True

2. $0 \leq p(x, y) \leq 1$ for all x.

ANSWER:

False

3. $0 \leq \int_{-\infty}^{\infty} p(x, y) \, dy \leq 1$ for all x.

ANSWER:

False

4. $\int_a^b p(x, y) \, dx$ is the probability that $a \leq x \leq b$.

ANSWER:

False

5. $p(a, b)$ is the probability that $x = a$ and $y = b$.

ANSWER:

False

6. $\int_{-\infty}^{\infty} \int_{-\infty}^{\infty} p(x, y) \, dy \, dx = 1$.

ANSWER:

True

7. $\lim_{T \to \infty} \int_{-\infty}^{\infty} \int_{-\infty}^{T} p(x, y) \, dy \, dx = 1$.

ANSWER:

True

ConcepTests and Answers and Comments for Section 16.7 ────────

In Problems 1–2, consider a change of variable in the integral $\int_R f(x, y)\, dA$ from x, y to s, t. Are the following statements true or false?

1. If the Jacobian $\left| \dfrac{\partial(x, y)}{\partial(s, t)} \right| > 1$, the value of the s, t-integral is greater than the original x, y-integral.

   ANSWER:

   False. The change of variable leaves the value of the integral unchanged. Thus, calculating the value of the s, t-integral gives you the value of the x, y-integral.

2. The Jacobian cannot be negative.

   ANSWER:

   False. The Jacobian can be negative; we use the absolute value of the Jacobian in the integral.

Chapter Seventeen

ConcepTests and Answers and Comments for Section 17.1

1. Which of the following parameterizations describes the quarter circle in Figure 17.1?

   (a) $(\cos t, \sin t)$, $0 \leq t \leq \frac{\pi}{2}$
   (c) $(-\cos t, \sin t)$, $\frac{\pi}{2} \leq t \leq \pi$

   (b) $(\sin t, \cos t)$, $0 \leq t \leq \frac{\pi}{2}$
   (d) $(\cos t, -\sin t)$, $\frac{3\pi}{2} \leq t \leq 2\pi$

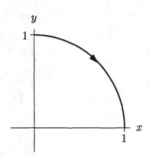

Figure 17.1

ANSWER:

(b), (c), (d). All except the first parameterization trace the given quarter circle.

COMMENT:

Students need practice with parameterizations of the unit circle, or any other curve. There are infinitely many different parameterizations of any given curve.

2. Let $(\cos at, \sin at)$ be the position at time t seconds of a particle moving around a circle, where $a > 0$. If a is increased,

   (a) The radius of the circle increases.
   (c) The center of the circle changes.

   (b) The speed of the particle increases.
   (d) The path ceases to be a circle.

   ANSWER:

   (b) The speed of the particle increases. As a increases, it takes less time for the particle to complete one revolution, thus the speed increases.

   COMMENT:

   Follow-up Questions. What has to happen to decrease the speed? How can you modify the formulas to change the radius and the center of the circle? To make the particle circle in the opposite direction?

3. Let $(a \cos t, a \sin t)$ be the position at time t seconds of a particle moving around a circle, where $a > 0$. If a is increased,

   (a) The radius of the circle increases.
   (c) The center of the circle changes.

   (b) The speed of the particle decreases.
   (d) The path ceases to be a circle.

   ANSWER:

   (a) The radius of the circle increases.

   COMMENT:

   Follow-up Questions. What has to happen to decreases the radius? How can you modify the parameterization to change the speed of the particle and the center of the circle?

4. Which of the following parametric curves trace out the unit circle (possibly more than once)?

   (a) $(\cos t, \sin t)$, $0 \leq t \leq 2\pi$
   (b) $(\sin^2 t, \cos^2 t)$, $0 \leq t \leq 2\pi$
   (c) $(\sin(t^2), \cos(t^2))$, $0 \leq t \leq 2\pi$
   (d) $(\sin 2t, \cos 2t)$, $0 \leq t \leq 2\pi$

   ANSWER:

   (a), (c), (d). All parametric curves except the second one trace the unit circle.

   COMMENT:

   Follow-up Questions. How do the parametric curves differ? How would a particle move along each of the curves? What does (b) look like?

5. Which of the following parametric paths describe particles that are traveling along a straight line in 3-space?

(a) $(1 - t, 2 + 2t, 3 - t)$
(b) $(1 - t^2, 2 + 2t^2, 3 - t^2)$
(c) $(1, 2, 1 - t)$
(d) $(1, t, 1 - t^2)$

ANSWER:

(a), (b), (c). All except the last particle move along straight lines.

COMMENT:

It is easy to think that if parameterization involves t^2, the path cannot be a line. But in (b) we could make a change of variable $s = t^2$ and see that the particle still moves along a straight line. For (b): What happens to the particle as t increases from negative to positive?

6. The value of c for which the lines $l(t) = (c + 4t, 2 - t, 3 + t)$ and $m(t) = (4t, 1 - 8t, 4 + 4t)$ intersect is

(a) 4
(b) 0
(c) −4
(d) There is no such c

ANSWER:

The two parameters, t, in $l(t)$ and $m(t)$, represent different quantities, so we will call the second one s. Thus $m(s) = (4s, 1 - 8s, 4 + 4s)$. The lines intersect if

$$c + 4t = 4s$$
$$2 - t = 1 - 8s$$
$$3 + t = 4 + 4s.$$

Solving the last two equations gives $t = 1$ and $s = 0$. Substituting into the first equation gives $c = -4$. Thus (c) is the correct answer.

COMMENT:

Two particles traveling along the lines with the given motions do not collide, but their paths cross if $c = -4$. **Follow-up Questions.** At what point in space do the paths cross? At what time does each of the particles reach the crossing?

ConcepTests and Answers and Comments for Section 17.2

1. Let $\vec{a}$ and $\vec{v}$ be the acceleration and velocity of a particle. Describe the motion of a particle such that

(a) $\vec{a}$ and $\vec{v}$ are parallel and point in the same direction.
(b) $\vec{a}$ and $\vec{v}$ are parallel and point in opposite directions.
(c) $\vec{a}$ and $\vec{v}$ are perpendicular.
(d) $\|\vec{a}\|$ and $\|\vec{v}\|$ are constant.

ANSWER:

Examples are:

(a) A particle moves along a straight line speeding up.
(b) A particle moves along a straight line slowing down.
(c) A particle moves around a circle at a constant speed.
(d) A particle moves around a circle at a constant speed.

Other answers are possible

COMMENT:

For part (d) the easiest situation is given when $\vec{a}$ and $\vec{v}$ are both the zero vector—the particle is not moving at all.

2. A car is moving on the race track in Figure 17.2. The car is going at constant speed as fast as possible without flying off the track. Let $\vec{a}$ be the acceleration of the car and $\vec{v}$ the velocity of the car. Mark points on the track where

   (a) $\vec{a} \cdot \vec{v} = 0$.
   (b) $\vec{a} \cdot \vec{v} > 0$.
   (c) $\vec{a} \cdot \vec{v} < 0$.

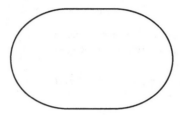

Figure 17.2

   ANSWER:

   (a) $\vec{a}$ and $\vec{v}$ are perpendicular at the leftmost and rightmost points on the track. At the midpoint of the straightaway, the acceleration is the zero vector.
   (b) Going into the straightaway (coming out of the tight curve), the angle between the two vectors is less than $\frac{\pi}{2}$.
   (c) Going into the tight curve the angle between the two vectors is larger than $\frac{\pi}{2}$.

   COMMENT:

   There are two very different sets of points where the dot product of the two vectors is zero. Discuss what is happening at those points. The acceleration vector is opposite the direction in which the driver feels pressed against the seat or doors of the race car. Some students may be able to solve this problem by vividly imagining themselves at the wheel during a race.

3. Are the following statements about a particle moving through space true or false? Give a counterexample to false statements.

   (a) If the velocity of a particle is zero, then its speed is zero.
   (b) If the speed of a particle is zero, its velocity must be zero.
   (c) A particle that is not accelerating must have zero velocity.
   (d) A particle with constant speed must have zero acceleration.
   (e) A particle with zero acceleration must have constant speed.
   (f) A particle with constant velocity must have zero acceleration.
   (g) A particle with constant speed must have constant velocity.

   ANSWER:

   (a) True
   (b) True
   (c) False: A particle moving with constant speed along a straight line is not accelerating.
   (d) False: A particle moving with constant speed around a circle does not have zero acceleration.
   (e) True
   (f) True
   (g) False: A particle moving around a circle with constant speed has changing velocity since its direction changes.

   COMMENT:

   If the particle is assumed to move in a straight line, some of the false statements would be true.

ConcepTests and Answers and Comments for Section 17.3

In Problems 1–2, match the formulas (a)-(d) with the graphs of vector fields (I)-(IV). You do not need to find formulas for f, g, h, and k.

1. Match the formulas (a)–(d) with the graphs of vector fields (I)–(IV).

   (a) $f(x)\vec{i}$
   (b) $g(x)\vec{j}$
   (c) $h(y)\vec{i}$
   (d) $k(y)\vec{j}$

(I)

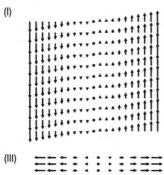

(II)

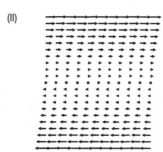

(III)

(IV)

ANSWER:

All vectors in the fields $f(x)\vec{i}$ and $h(y)\vec{i}$ point in the horizontal direction, since they are all multiples of the horizontal vector $\vec{i}$. The graphs in (II) and (III) are the ones with horizontal vectors, but which is (a) and which is (c)? Since the length of the vectors in (II) changes when y changes (moving upward in the graph) but not when x changes (moving horizontally), graph (II) must correspond to (c), $h(y)\vec{i}$. Thus (III) goes with (a), $f(x)\vec{i}$.

All vectors in the fields $g(x)\vec{j}$ and $k(y)\vec{j}$ point in the vertical direction, since they are all multiples of the vertical vector $\vec{j}$. The graphs in (I) and (IV) are the ones with vertical vectors. The length of the vectors in (I) changes when x changes but not when y changes, so graph (I) must correspond to (b), $g(x)\vec{j}$. Thus (IV) goes with (d), $k(y)\vec{j}$.

The match-up is: a-III, b-I, c-II, d-IV.

COMMENT:

Students should be encouraged to read formulas for qualitative features.

2. Match the formulas (a)–(d) with the graphs of vector fields (I)–(IV).

   (a) $f(x)\vec{i} + f(x)\vec{j}$
   (b) $g(x)\vec{i} - g(x)\vec{j}$
   (c) $h(y)\vec{i} + h(y)\vec{j}$
   (d) $k(y)\vec{i} - k(y)\vec{j}$

(I)
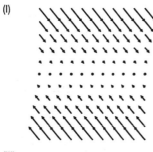

(II)
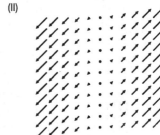

(III)
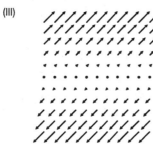

(IV)
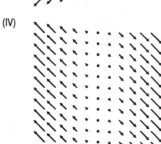

ANSWER:

All vectors in the fields $f(x)\vec{i} + f(x)\vec{j} = f(x)(\vec{i} + \vec{j})$ and $h(y)\vec{i} + h(y)\vec{j} = h(y)(\vec{i} + \vec{j})$ point parallel to $\vec{i} + \vec{j}$. The graphs in (II) and (III) are the ones with vectors parallel to $\vec{i} + \vec{j}$, but which is (a) and which is (c)? Since the length of the vectors in (II) changes when x changes (moving horizontally in the graph) but not when y changes (moving vertically), graph (II) must correspond to (a), $f(x)\vec{i} + f(x)\vec{j}$. Thus (III) goes with (c), $h(y)\vec{i} + h(y)\vec{j}$.

All vectors in the fields $g(x)\vec{i} - g(x)\vec{j} = g(x)(\vec{i} - \vec{j})$ and $k(y)\vec{i} - k(y)\vec{j} = k(y)(\vec{i} - \vec{j})$ point parallel to $\vec{i} - \vec{j}$. The graphs in (I) and (IV) are the ones in this direction. The length of the vectors in (I) changes when y changes but not when x changes, so graph (I) must correspond to (d), $k(y)\vec{i} - k(y)\vec{j}$. Thus (IV) goes with (b), $g(x)\vec{i} - g(x)\vec{j}$.

The match-up is: a-II, b-IV, c-III, d-I.

3. Match the vector fields with the appropriate pictures.

   (a) $\vec{F}_1(\vec{r}) = \frac{\vec{r}}{||\vec{r}||}$

   (b) $\vec{F}_2(\vec{r}) = \vec{r}$

   (c) $\vec{F}_3(x,y) = y\vec{i} - x\vec{j}$

   (d) $\vec{F}_4(x,y) = x\vec{j}$

(i)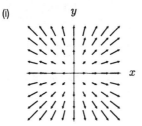

(ii)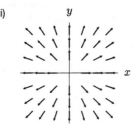

(iii)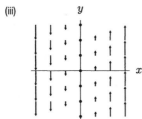

(iv)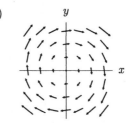

ANSWER:

(a) and (ii), (b) and (i), (c) and (iv), (d) and (iii)

COMMENT:

Process of elimination helps to do this problem.

4. For each of the vector fields in the plane, $\vec{F}_1, \vec{F}_2, \vec{F}_3$, decide which of the properties the vector field has:

$$\vec{F}_1(\vec{r}) = \frac{\vec{r}}{||\vec{r}||} \qquad \vec{F}_2(\vec{r}) = -\vec{r} \qquad \vec{F}_3(x,y) = \frac{\vec{i} + \vec{j}}{x^2 + y^2}$$

A All vectors point toward the origin.

B All vectors point away from the origin.

C All vectors point in the same direction.

D All vectors have the same length.

E Vectors get longer away from the origin.

F Vectors get shorter away from the origin.

G The vector field has rotational symmetry about the origin.

H Vectors that start on the same circle centered at the origin have the same length.

   ANSWER:

   $\vec{F}_1$: B, D, G, H

   $\vec{F}_2$: A, E, G, H

   $\vec{F}_3$: C, F, H

   COMMENT:

   $\vec{F}_3$ is the most interesting of the vector fields. All vectors points northeast, and vectors on the same circle have constant length.

5. Compare and contrast the following vector fields:

$$\vec{F}_1(\vec{r}) = \frac{\vec{r}}{||\vec{r}||} \qquad \vec{F}_2(\vec{r}) = \vec{r} \qquad \vec{F}_3(\vec{r}) = -\frac{\vec{r}}{||\vec{r}||^2}$$

ANSWER:
$\vec{F}_1$ and $\vec{F}_2$ look very similar, vectors point away from the origin. The only difference is that in $\vec{F}_1$ all vectors have length 1, and in $\vec{F}_2$ vectors get longer as they move away from the origin. In $\vec{F}_3$ vectors point toward the origin and get shorter the farther away from the origin they are.

COMMENT:
Vector fields of this form are used frequently for line and flux integrals. It is an advantage to be able to visualize them.

6. Figure 17.3 shows the vector field $\vec{F} = \text{grad } f$. Which of the following are possible choices for $f(x, y)$?

(a) x^2

(b) $-x^2$

(c) $-2x$

(d) $-y^2$

Figure 17.3

ANSWER:
(b), $-x^2$, since $\text{grad}(-x^2) = -2x\vec{i}$, which is parallel to the x axis and points toward the y-axis. (a) is not possible because its gradient points away from the y-axis, and (c) is not possible because its gradient is the constant vector $-2\vec{i}$. (d) is not possible because its gradient is parallel to the y-axis.

COMMENT:
Ask students for other possible functions whose gradient is the vector field shown.

7. Rank the length of the gradient vectors at the points marked on the contour plot in Figure 17.4.

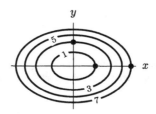

Figure 17.4

ANSWER:
All vectors point radially outward. Vectors on the contour $z = 7$ are longer than vectors on the contour $z = 3$ which are longer than vectors on the contour $z = 1$.

COMMENT:
While the exact direction and length of the vectors is not that important in this problem, a qualitative comparison of their lengths is.

ConcepTests and Answers and Comments for Section 17.4

1. Which parameterized curve is not a flow line of the vector field $\vec{F} = x\vec{i} + y\vec{j}$?

   (a) $\vec{r}(t) = e^t\vec{i} + e^t\vec{j}$
   (b) $\vec{r}(t) = e^t\vec{i} + 2e^t\vec{j}$
   (c) $\vec{r}(t) = 3e^t\vec{i} + 3e^t\vec{j}$
   (d) $\vec{r}(t) = 2e^t\vec{i} + e^{2t}\vec{j}$

   ANSWER:
   (d) The flow lines $\vec{r}(t) = x(t)\vec{i} + y(t)\vec{j}$ solve the differential equations $dx/dt = x$, $dy/dt = y$. The solutions are all of the form $x = ae^t$ and $y = be^t$, so all but (d) are flow lines.

   COMMENT:
   How do the flow lines (a), (b), (c) for the vector field $\vec{F}$ differ? Ask the students to describe the parameterized curve in (d) (parabolic, on the parabola $y = x^2/4$) or to find a vector field of which it is a flow line (for example, $\vec{F}_1 = x\vec{i} + 2y\vec{j}$).

2. Which parameterized curves are flow lines of the vector field $\vec{F} = -y\vec{i} + x\vec{j}$?

   (a) $\vec{r}(t) = \cos t\vec{i} + \sin t\vec{j}$
   (b) $\vec{r}(t) = \cos t\vec{i} - \sin t\vec{j}$
   (c) $\vec{r}(t) = \sin t\vec{i} - \cos t$
   (d) $\vec{r}(t) = 2\cos t\vec{i} + 2\sin t\vec{j}$

   ANSWER:
   (a), (c), and (d) The flow lines $\vec{r}(t) = x(t)\vec{i} + y(t)\vec{j}$ solve the system of differential equations $dx/dt = -y$, $dy/dt = x$. A direct check shows that only (b) fails to yield a solution.

   COMMENT:
   How do the flow lines (a), (c), and (d) for the vector field $\vec{F}$ differ? Ask the students to describe the parameterized curve in (b) (motion clockwise around a circle) or to find a vector field of which it is a flow line (for example, $\vec{F}_1 = y\vec{i} - x\vec{j}$).

3. Match the vector fields (a)–(d) and flow lines I–IV.

(a) $\vec{i} + y\vec{j}$	I. $x = t, y = t^3/3$
(b) $\vec{i} + x^2\vec{j}$	II. $x = t^2/2, y = t$
(c) $x\vec{i} + x\vec{j}$	III. $x = e^t, y = e^t$
(d) $y\vec{i} + \vec{j}$	IV. $x = t, y = e^t$

   ANSWER:
   Match by substituting I–IV into $\vec{r}'(t) = \vec{F}(\vec{r}(t))$, where $\vec{F}$ is each of the vector fields; alternatively, solve the differential equations. Then

   (a) IV   (b) I   (c) III   (d) II

4. An object flowing in a vector field is at the point $(1, 2)$ at time $t = 5.00$. Using one step of Euler's method, match the velocity fields $\vec{F}_i$ in (I)–(IV) with the approximate position of the object at time $t = 5.01$ given in (a)–(d).

<table>
<tr><td>(a) $(1.02, 2)$</td><td>I. $\vec{F}_1 = x\vec{i} + y\vec{j}$</td></tr>
<tr><td>(b) $(1.02, 2.01)$</td><td>II. $\vec{F}_2 = y\vec{i} + x\vec{j}$</td></tr>
<tr><td>(c) $(1.01, 2)$</td><td>III. $\vec{F}_3 = x\vec{i}$</td></tr>
<tr><td>(d) $(1.01, 2.02)$</td><td>IV. $\vec{F}_4 = y\vec{i}$</td></tr>
</table>

ANSWER:

Writing $\vec{r}(t)$ for the position vector of the object at time t, we have

$$\vec{r}(5.01) = \vec{r}(5) + 0.01\vec{F}_i(1, 2) = (\vec{i} + 2\vec{j}) + 0.01\vec{F}_i(1, 2).$$

(I) We have $\vec{r}(5.01) = (\vec{i} + 2\vec{j}) + 0.01(\vec{i} + 2\vec{j}) = 1.01\vec{i} + 2.02\vec{j}$. At $t = 5.01$ the object is at the point $(1.01, 2.02)$.

(II) We have $\vec{r}(5.01) = (\vec{i} + 2\vec{j}) + 0.01(2\vec{i} + \vec{j}) = 1.02\vec{i} + 2.01\vec{j}$. At $t = 5.01$ the object is at the point $(1.02, 2.01)$.

(III) We have $\vec{r}(5.01) = (\vec{i} + 2\vec{j}) + 0.01\vec{i} = 1.01\vec{i} + 2\vec{j}$. At $t = 5.01$ the object is at the point $(1.01, 2.00)$.

(IV) We have $\vec{r}(5.01) = (\vec{i} + 2\vec{j}) + 0.01(2\vec{i}) = 1.02\vec{i} + 2\vec{j}$. At $t = 5.01$ the object is at the point $(1.02, 2.00)$.

The match-up is: I-d, II-b, III-c, IV-a

COMMENT:

Ask the students to approximate the position of the particle at $t = 5.02$ by completing a second step of Euler's method with $\Delta t = 0.01$.

5. True or false?

(a) Two different curves can be flow lines for the same vector field.

(b) If one parameterization of a curve is a flow line for a vector field, then all its parameterizations are flow lines for the vector field.

ANSWER:

(a) True. Every vector field has many flow lines. There is a flow line for a given vector field through every point in the plane.

(b) False. A flow line of a velocity field represents the motion of a particle with velocity specified by the vector field at every point. Changing the parameterization can change the velocity of the particle even though the particle moves through exactly the same points.

COMMENT:

Ask the students to give examples.

6. Are the following true or false? If $\vec{r}(t)$ is a flow line for a vector field $\vec{F}$, then

(a) $\vec{r}_1(t) = \vec{r}(t - 5)$ is a flow line for the same vector field $\vec{F}$.

(b) $\vec{r}_1(t) = \vec{r}(2t)$ is a flow line for the vector field $2\vec{F}$.

(c) $\vec{r}_1(t) = 2\vec{r}(t)$ is a flow line for the vector field $2\vec{F}$.

ANSWER:

(a) True. Two particles with motion described by $\vec{r}(t)$ and $\vec{r}_1(t) = \vec{r}(t - 5)$ move along exactly the same path, with the second particle following 5 time units behind. Their velocities at a given point are the same and so can be given by the same velocity field $\vec{F}$.

(b) True. Two particles with motion described by $\vec{r}(t)$ and $\vec{r}_1(t) = \vec{r}(2t)$ move along exactly the same path, with the second particle moving twice as fast as the first at any given point. The velocity of the first at a point on its path is $\vec{F}$, so the velocity of the second is $2\vec{F}$.

(c) False. A counterexample is given by $\vec{F} = x\vec{i}$, and $\vec{r}(t) = e^t\vec{i}$. Then $\vec{r}_1(t) = 2e^t\vec{i}$ is not a flow line of $2\vec{F} = 2x\vec{i}$.

ConcepTests and Answers and Comments for Section 17.5

1. Match the statements (a)–(e) and surfaces I–V. Note that r, θ, z are cylindrical coordinates; ρ, ϕ, θ are spherical coordinates.

   (a) $0 \leq r \leq 5, z = 10$ I.   Cylinder
   (b) $0 \leq z \leq 10, r = 5$ II.   Cone
   (c) $0 \leq \rho \leq 5, \phi = \pi/3$ III.   Disk above xy-plane
   (d) $0 \leq \rho \leq 5, \phi = \pi/2$ IV.   Half plane
   (e) $\theta = \pi/2$ V.   Disk in xy-plane

   ANSWER:

   (a)   III   (b)   I   (c)   II   (d)   V   (e)   IV

   COMMENT:

   Ask the students to give parameterizations for the five surfaces. Point out that they need two and not three parameters and that they cannot use the two coordinates appearing in statements (a)–(e) for parameters.

2. Match the statements (a)–(e) and parameterizations I–V. Note that r, θ, z are cylindrical coordinates; ρ, ϕ, θ are spherical coordinates.

   (a)   Disk above xy-plane   I. $x = 0, y = \rho \sin \phi, z = \rho \cos \phi, 0 \leq \rho < \infty, 0 \leq \phi \leq \pi$
   (b)   Cylinder   II. $x = \rho \sin(\pi/3) \cos \theta, y = \rho \sin(\pi/3) \sin \theta, z = \rho \cos(\pi/3),$
   (c)   Cone $0 \leq \theta \leq 2\pi, 0 \leq \rho \leq 5$
   (d)   Disk in the xy-plane   III. $x = r \cos \theta, y = r \sin \theta, z = 10, 0 \leq r \leq 5, 0 \leq \theta \leq 2\pi$
   (e)   Half-plane   IV. $x = 5 \cos \theta, y = 5 \sin \theta, z = t, 0 \leq \theta \leq 2\pi, 0 \leq t \leq 10$
                                   V. $x = \rho \cos \theta, y = \rho \sin \theta, z = 0, 0 \leq \theta \leq 2\pi, 0 \leq \rho \leq 5$

   ANSWER:

   (a)   III   (b)   IV   (c)   II   (d)   V   (e)   I

   COMMENT:

   Ask the students to give an r, θ parameterization of (e) in place of the ρ, ϕ parameterization I.

3. Match the statements (a)–(e) and parameterizations I–V. Note that r, θ, z are cylindrical coordinates; ρ, ϕ, θ are spherical coordinates.

   (a) $0 \leq r \leq 5, z = 10$ I. $x = 0, y = \rho \sin \phi, z = \rho \cos \phi, 0 \leq \rho < \infty, 0 \leq \phi \leq \pi$
   (b) $0 \leq z \leq 10, r = 5$ II. $x = \rho \sin(\pi/3) \cos \theta, y = \rho \sin(\pi/3) \sin \theta, z = \rho \cos(\pi/3),$
   (c) $0 \leq \rho \leq 5, \phi = \pi/3$ $0 \leq \theta \leq 2\pi, 0 \leq \rho \leq 5$
   (d) $0 \leq \rho \leq 5, \phi = \pi/2$ III. $x = r \cos \theta, y = r \sin \theta, z = 10, 0 \leq r \leq 5, 0 \leq \theta \leq 2\pi$
   (e) $\theta = \pi/2$ IV. $x = 5 \cos \theta, y = 5 \sin \theta, z = t, 0 \leq \theta \leq 2\pi, 0 \leq t \leq 10$
                                   V. $x = \rho \cos \theta, y = \rho \sin \theta, z = 0, 0 \leq \theta \leq 2\pi, 0 \leq \rho \leq 5$

   ANSWER:

   (a)   III   (b)   IV   (c)   II   (d)   V   (e)   I

   COMMENT:

   Ask the students to give an r, θ parameterization of (e) in place of the ρ, ϕ parameterization I.

4. A hemisphere of the sphere of radius 5 centered at the origin is parameterized by spherical coordinates in the usual way:

$$x = 5 \sin \phi \cos \theta \qquad y = 5 \sin \phi \sin \theta \qquad z = 5 \cos \phi.$$

Match the hemisphere (a)–(d) with the parameter limits (I)–(IV).

   (a) $z \geq 0$ I. $0 \leq \theta \leq \pi, \;\; 0 \leq \phi \leq \pi$
   (b) $y \geq 0$ II. $\pi/2 \leq \theta \leq 3\pi/2, \;\; 0 \leq \phi \leq \pi$
   (c) $x \leq 0$ III. $0 \leq \theta \leq 2\pi, \;\; 0 \leq \phi \leq \pi/2$
   (d) $y \geq x$ IV. $\pi/4 \leq \theta \leq 5\pi/4, \;\; 0 \leq \phi \leq \pi$

   ANSWER:

   The match-up is: I-b, II-c, III-a, IV-d.

   COMMENT:

   Ask the students to find parameter limits for other hemispheres.

5. Match the parameterizations (I)–(IV) of the cylinders with their length L and radius R in (a)–(d). In all cases $0 \leq s \leq 2\pi$, $0 \leq t \leq 1$.

I.	$x = 5\cos s$	$y = t$	$z = 5\sin s$	(a)	$L = 1, R = 1$
II.	$x = 5\cos s$	$y = 5t$	$z = 5\sin s$	(b)	$L = 1, R = 5$
III.	$x = \cos s$	$y = t$	$z = \sin s$	(c)	$L = 5, R = 1$
IV.	$x = \cos s$	$y = 5t$	$z = \sin s$	(d)	$L = 5, R = 5$

ANSWER:

I-b, II-d, III-a, IV-c.

COMMENT:

What about $x = \cos 5s \qquad y = t \qquad z = \sin 5s$?

6. Match the parameterizations (I)–(IV) with the surfaces (a)–(d). In all cases $0 \leq s \leq \pi/2, 0 \leq t \leq \pi/2$. Note that only part of the surface may be described by the given parameterization.

(a) Cylinder	I. $x = \cos s, \quad y = \sin t, \quad z = \cos s + \sin t$
(b) Plane	II. $x = \cos s, \quad y = \sin s, \quad z = \cos t$
(c) Sphere	III. $x = \sin s \cos t, \quad y = \sin s \sin t, \quad z = \cos s$
(d) Cone	IV. $x = \cos s, \quad y = \sin t, \quad z = \sqrt{\cos^2 s + \sin^2 t}$

ANSWER:

(I) Part of a plane, since $z = x + y$.

(II) Part of a cylinder of radius 1 with center on the z-axis, since $x^2 + y^2 = 1$.

(III) Part of a sphere of radius 1 centered at the origin, since $x^2 + y^2 + z^2 = 1$. With $s = \phi$ and $t = \theta$ this is the usual parameterization of the sphere using spherical coordinates.

(IV) Part of a cone with vertex at the origin and central axis on the z-axis, since $z = \sqrt{x^2 + y^2}$.

The match-up is: I-b, II-a, III-c, IV-d

COMMENT:

Ask what part of each surface is described with $0 \leq s, t \leq \pi/2$.

7. True or false? The parameter curves on a parameterized plane are straight lines.

ANSWER:

False. The parameter curves depend on the parameterization. They could be straight lines, but they don't have to be. For example, if the xy-plane is parameterized in the usual way by polar coordinates, $x = r\cos\theta$, $y = r\sin\theta$, $z = 0$, then the parameter curves with r constant are circles.

COMMENT:

Ask the students if there is a parameterization of the plane with parabolas as parameter curves.

Chapter Eighteen

ConcepTests and Answers and Comments for Section 18.1

1. The vector field $\vec{F}$ and the oriented curves C_1, C_2, C_3, C_4 are shown in Figure 18.1. For which of the paths C_i is the line integral $\int_{C_i} \vec{F} \cdot d\vec{r}$ positive?

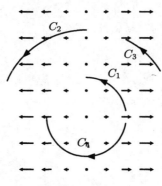

Figure 18.1

ANSWER:

C_2. The vector field $\vec{F}$ and the line segments $\Delta \vec{r}$ are approximately in opposite directions for the curves C_1 and C_3. So the contributions of each term $\vec{F} \cdot \Delta \vec{r}$ are negative for these curves. Thus, $\int_{C_1} \vec{F} \cdot d\vec{r}$ and $\int_{C_3} \vec{F} \cdot d\vec{r}$ are negative. For the curve C_4, the contribution from the first half of the curve cancels out the contribution from the second half, so $\int_{C_4} \vec{F} \cdot d\vec{r} = 0$. For the curve C_2, the vector field and line segment $\Delta \vec{r}$ are approximately in the same direction, so $\int_{C_2} \vec{F} \cdot d\vec{r} > 0$.

COMMENT:

Point out that symmetry arguments can be used to show an integral is zero. Illustrate with a path symmetric about the x axis.

2. Given C, a circle centered at the origin oriented clockwise, say for which of the vector fields below the circulation around C is positive.

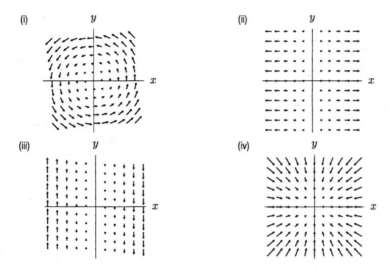

Figure 18.2

ANSWER:

(iii). The vector field in (i) circles counterclockwise around the origin, so its circulation around C is negative. For the vector field in (ii), the path integral along the left half of C cancels out the path integral along the right half, so the circulation is zero. The vector field in (iii) has a component in the direction of C all the way around (except at the very top and bottom of C), so the circulation is positive. The vector field in (iv) is perpendicular to C at every point, so the circulation is zero.

COMMENT:

For the vector field in (ii) you can also use a symmetry argument about the x-axis. In fact, the integral along each of the top, bottom, left, or right semicircles is zero.

3. True or false? Given two circles centered at the origin, oriented counterclockwise, and any vector field $\vec{F}$, then the path integral of $\vec{F}$ is larger around the circle with larger radius.

ANSWER:

False. The path integral is determined by the values of the vector field along the path, not simply by the size of the path.

COMMENT:

For example, the vector field $-y/(x^2 + y^2)^{3/2}\vec{i} + x/(x^2 + y^2)^{3/2}\vec{j}$ has a larger path integral around the circle of radius 1 than around the circle of radius 2.

4. True or false? If $\vec{F}$ is any vector field and C is a circle, then the integral of $\vec{F}$ around C traversed clockwise is the negative of the integral of $\vec{F}$ around C traversed counterclockwise.

ANSWER:

True. The integral of a vector field along a path changes sign when the path is reversed.

COMMENT:

Note that the statement is true even when the integral is zero, since in that case both integrals are zero, and the negative of zero is zero.

5. Group together the equal line integrals $\int_C \vec{F} \cdot d\vec{r}$ in (I)–(IV), where C is the oriented curve from P to Q, and $\vec{F}$ is the vector field shown. The scale is the same on all four graphs.

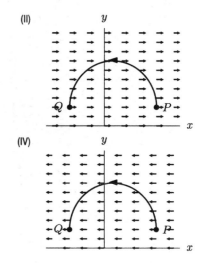

ANSWER:

The vector fields in (I) and (II) look the same, but the curves are oriented in opposite directions, so the line integrals in (I) and (II) have approximately the same absolute value but opposite sign.

The oriented curves in (I) and (III) look the same but the vector fields point in opposite directions (with the same magnitudes), so the line integrals in (I) and (III) have the same absolute value but opposite sign. Thus (II) and (III) give approximately equal line integrals, both opposite to (I).

The vector fields in (III) and (IV) look the same, but the curves are oriented in opposite directions, so the line integrals in (III) and (IV) have approximately the same absolute value but opposite sign. Thus (I) and (IV) give equal line integrals.

The grouping by equality is: (I) with (IV); (II) with (III).

6. Are the following statements about a line integral $\int_C \vec{F} \cdot d\vec{r}$ true or false? If the angle between the vector field $\vec{F}$ and the oriented curve C at every point of C is

(a) An acute angle, then $\int_C \vec{F} \cdot d\vec{r} > 0$.

(b) A right angle, then $\int_C \vec{F} \cdot d\vec{r} = 0$.

(c) An obtuse angle, then $\int_C \vec{F} \cdot d\vec{r} < 0$.

ANSWER:

The line integrals $\int_C \vec{F} \cdot d\vec{r}$ are approximated by Riemann sums

$$\sum_i \vec{F}(\vec{r}_i) \cdot \Delta\vec{r}_i = \sum_i \|\vec{F}(\vec{r}_i)\|\|\Delta\vec{r}_i\| \cos\theta_i$$

where θ_i is (approximately) the angle between the vector field $\vec{F}$ (represented by the vector $\vec{F}(\vec{r}_i)$) and the oriented curve C (approximated by the vector $\Delta\vec{r}_i$)

(a) True. In the Riemann sums all dot product terms are positive because the angle θ_i is acute, giving $\cos\theta_i > 0$.

(b) True. In the limit, all dot product terms are zero because the angle θ_i is a right angle, giving $\cos\theta_i = 0$.

(c) True. In the Riemann sums all dot product terms are negative because the angle θ_i is obtuse, giving $\cos\theta_i < 0$.

7. The work done by the force field $\vec{F} = y\vec{i}$ as an object moves along a straight line joining $(1, 1)$ to $(1, -1)$ is

   (a) positive
   (b) negative
   (c) zero.

   ANSWER:

   (c). Since the two points have the same x-coordinate, the straight line joining them is vertical. On the other hand, the vector field points in the horizontal direction, and so is always perpendicular to the path of the object. Thus it does not work on the object as it moves.

   COMMENT:

   Follow-up. Ask the students to give paths for which the work is positive or negative. Ask for a vector field that would make the work positive or negative along the given path.

8. A Riemann sum approximation for a line integral $\int_C \text{grad } f \cdot d\vec{r} \approx \Sigma \text{grad } f(\vec{r}_i) \cdot \Delta \vec{r}_i$ is given for a very fine subdivision of the oriented curve C. Which of the following can you conclude? The terms of the Riemann sum coming from segments of C

   (a) Where C crosses level curves of f in the direction of increasing values are positive.
   (b) Where f is positive are positive.
   (c) Where grad f is positive are positive
   (d) Where C is tangent to a level curve of f are positive.

   ANSWER:

   (a) The vector grad $f(\vec{r}_i)$ is perpendicular to the level curve of f at a point P of C, and it points in the direction of increasing values of f. If C crosses the level curve at P in the direction of increasing values of f, then a tangent vector to C at P in the orientation direction forms an acute angle with grad $f(\vec{r}_i)$. The vector $\Delta \vec{r}_i$ is very nearly tangent to C, so it, too, will form an acute angle with grad $f(\vec{r}_i)$. Thus grad $f(\vec{r}_i) \cdot \Delta \vec{r}_i > 0$, a positive term in the Riemann sum.

   Counterexamples can be constructed for statements (b) and (d). Statement (c) does not make sense because grad f is a vector, and a vector can not be positive or negative, only a scalar can. Thus, statement (a) is the only conclusion we can make.

ConcepTests and Answers and Comments for Section 18.2

1. If C_1 is the path parameterized by $\vec{r}_1(t) = (t, t)$, $0 \le t \le 1$, and if C_2 is the path parameterized by $\vec{r}_2(t) = (t^2, t^2)$, $0 \le t \le 1$, and if $\vec{F} = x\vec{i} + y\vec{j}$, which of the following is true?

   (a) $\int_{C_1} \vec{F} \cdot d\vec{r} > \int_{C_2} \vec{F} \cdot d\vec{r}$
   (b) $\int_{C_1} \vec{F} \cdot d\vec{r} < \int_{C_2} \vec{F} \cdot d\vec{r}$
   (c) $\int_{C_1} \vec{F} \cdot d\vec{r} = \int_{C_2} \vec{F} \cdot d\vec{r}$

   ANSWER:

   (c) The two parameterizations give the same path, in the same direction. Since a line integral does not depend on the way the path is parameterized, the two line integrals are the same.

   COMMENT:

   Have the students calculate the integrals explicitly to show the equality.

2. If C_1 is the path parameterized by $\vec{r}_1(t) = (t, t)$, $0 \le t \le 1$, and if C_2 is the path parameterized by $\vec{r}_2(t) = (1-t, 1-t)$, $0 \le t \le 1$, and if $\vec{F} = x\vec{i} + y\vec{j}$, which of the following is true?

   (a) $\int_{C_1} \vec{F} \cdot d\vec{r} > \int_{C_2} \vec{F} \cdot d\vec{r}$
   (b) $\int_{C_1} \vec{F} \cdot d\vec{r} < \int_{C_2} \vec{F} \cdot d\vec{r}$
   (c) $\int_{C_1} \vec{F} \cdot d\vec{r} = \int_{C_2} \vec{F} \cdot d\vec{r}$

   ANSWER:

   (a). The two parameterizations give the same path, but one goes in the opposite direction to the other. Since $\vec{F}$ points away from the origin, and C_1 is oriented away from the origin, $\int_{C_1} \vec{F} \cdot d\vec{r}$ is positive, so $\int_{C_2} \vec{F} \cdot d\vec{r}$ is negative.

   COMMENT:

   Point out to students that when comparing two paths to see if they are the same, you have to compare their orientations as well as their shape.

3. If C_1 is the path parameterized by $\vec{r}_1(t) = (t, t)$, for $0 \le t \le 1$, and if C_2 is the path parameterized by $\vec{r}_2(t) = (\sin t, \sin t)$, for $0 \le t \le 1$, and if $\vec{F} = x\vec{i} + y\vec{j}$, which of the following is true?

(a) $\int_{C_1} \vec{F} \cdot d\vec{r} > \int_{C_2} \vec{F} \cdot d\vec{r}$

(b) $\int_{C_1} \vec{F} \cdot d\vec{r} < \int_{C_2} \vec{F} \cdot d\vec{r}$

(c) $\int_{C_1} \vec{F} \cdot d\vec{r} = \int_{C_2} \vec{F} \cdot d\vec{r}$

ANSWER:

(a). The two parameterizations move along the line $y = x$ in the direction away from the origin, but they have different endpoints. The path C_1 goes from $(0, 0)$ to $(1, 1)$ and the path C_2 goes from $(0, 0)$ to $(\sin 1, \sin 1) \approx (0.84, 0.84)$. Since $\vec{F}$ points away from the origin, and both paths are oriented so that the direction of travel is oriented away from the origin, the integral along the longer path is larger, so $\int_{C_1} \vec{F} \cdot d\vec{r} > \int_{C_2} \vec{F} \cdot d\vec{r}$.

COMMENT:

Point out to students that for two parameterizations to give the same path, they have to have the same endpoints.

4. Consider the path C_1 parameterized by $\vec{r}_1(t) = (\cos t, \sin t)$, $0 \le t \le 2\pi$ and the path C_2 parameterized by $\vec{r}_2(t) = (2\cos t, 2\sin t)$, $0 \le t \le 2\pi$. Let $\vec{F}$ be a vector field. Is it always true that $\int_{C_2} \vec{F} \cdot d\vec{r} = 2\int_{C_1} \vec{F} \cdot d\vec{r}$?

ANSWER:

No. C_1 is a circle of radius 1 centered at the origin and C_2 is a circle of radius 2 centered at the origin. There is no reason why the integral around a circle of radius 2 should be twice the size of the integral around a circle of radius 1. The integral depends on the values of the vector field along the path, not just on the shape of the path.

COMMENT:

For example, the vector field $-y/(x^2 + y^2)\vec{i} + x/(x^2 + y^2)\vec{j}$ has the same integral around both paths.

5. Consider the path C_1 parameterized by $\vec{r}_1(t) = (\cos t, \sin t)$, $0 \le t \le 2\pi$ and the path C_2 parameterized by $\vec{r}_2(t) = (\cos(2t), \sin(2t))$, $0 \le t \le 2\pi$. Let $\vec{F}$ be a vector field. Is it always true that $\int_{C_2} \vec{F} \cdot d\vec{r} = 2\int_{C_1} \vec{F} \cdot d\vec{r}$?

ANSWER:

Yes. C_1 is a circle of radius 1 centered at the origin, traversed once in the counterclockwise direction, and C_2 is the same circle traversed twice. Thus the integral around C_2 will be twice the integral around C_1.

COMMENT:

Follow-up. Ask the students to give a parameterized path C_3 that will make $\int_{C_3} \vec{F} \cdot d\vec{r} = -2\int_{C_1} \vec{F} \cdot d\vec{r}$.

ConcepTests and Answers and Comments for Section 18.3

1. The line integral of $\vec{F} = \operatorname{grad} f$ along one of the paths shown in Figure 18.3 is different from the integral along the other two. Which is the odd one out?

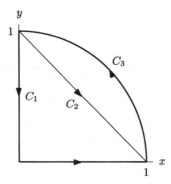

Figure 18.3

ANSWER:

C_3. The two paths C_1 and C_2 give the same line integral by the Fundamental Theorem of Calculus for line integrals, since they have the same starting and ending points. The integral over C_3 could be different because it goes in the opposite direction, so the starting and ending points are swapped.

COMMENT:

Follow-up Question. Is it possible for all integrals to be the same? Yes, if $f(0, 1) = f(1, 0)$, then all three integrals are zero.

2. Rank the four line integrals $\int_{C_i} \operatorname{grad} f \cdot d\vec{r}$ from smallest to largest, where C_i, $i = 1, 2, 3, 4$ are the oriented curves on the contour diagram of f in Figure 18.4.

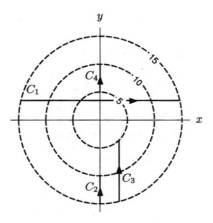

Figure 18.4

ANSWER:
By the Fundamental Theorem of Calculus for Line Integrals,

$$\int_{C_i} \operatorname{grad} f \cdot d\vec{r} = f(\text{point at end of } C_i) - f(\text{point at start of } C_i),$$

the difference of the values on the contour lines of f where the curve C_i starts and ends. Thus:

$I_1 = \int_{C_1} \operatorname{grad} f \cdot d\vec{r} = 15 - 15 = 0.$

$I_2 = \int_{C_2} \operatorname{grad} f \cdot d\vec{r} = 10 - 15 = -5.$

$I_3 = \int_{C_3} \operatorname{grad} f \cdot d\vec{r} = 5 - 15 = -10$

$I_4 = \int_{C_4} \operatorname{grad} f \cdot d\vec{r} = 10 - 5 = 5.$

The ranking is $I_3 < I_2 < I_1 < I_4$.

3. Figure 18.5 shows the vector field grad f, where f is continuously differentiable in the whole plane. The two ends of an oriented curve C from P to Q are shown, but the middle portion of the path is outside the viewing window pictured in Figure 18.5. The line integral $\int_C \text{grad } f \cdot d\vec{r}$ is

(a) Positive
(b) Negative
(c) Zero
(d) Can't tell without further information

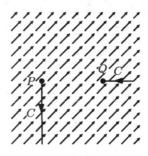

Figure 18.5

ANSWER:

(a) The line integral of the continuous gradient grad f is path independent, so $\int_C \text{grad } f \cdot d\vec{r} = \int_{C_1} \text{grad } f \cdot d\vec{r}$ where C_1 is the oriented straight-line path from P to Q in Figure 18.6. Since grad f forms an acute angle with the direction of C_1 at every point, the line integral over C_1 is positive. Thus, the line integral over C is also positive.

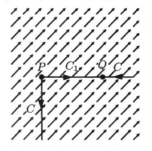

Figure 18.6

4. Which of the diagrams contain all three of the following: a contour diagram of a function f, the vector field grad f of the same function, and an oriented path C from P to Q with $\int_C$ grad $f \cdot d\vec{r} = 60$?

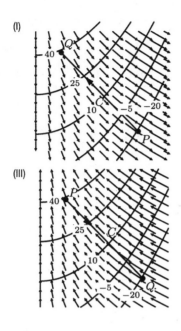

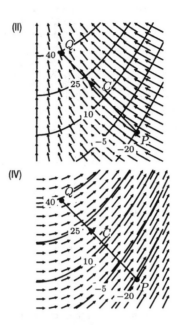

ANSWER:

The vector field grad f is perpendicular to the level curves of f, so (IV) is wrong. Moreover, grad f points in the direction of increasing values of f, so (I) is wrong. Finally, by the Fundamental Theorem of Line Integrals, $\int_C$ grad $f \cdot d\vec{r} = f(Q) - f(P)$, where $f(P)$ is the value of f at the starting point of C and $f(Q)$ is the value of f at the terminating point of C. In (II), $f(Q) - f(P) = 40 - (-20) = 60$ and in (III) $f(Q) - f(P) = -20 - 40 = -60$. The correct answer is (II)

5. If f is a smooth function of two variables that is positive everywhere and $\vec{F} =$ grad f, which of the following can you conclude about $\int_C \vec{F} \cdot d\vec{r}$?

   (a) It is positive for all smooth paths C
   (b) It is zero for all smooth paths C
   (c) It is positive for all closed smooth paths C.
   (d) It is zero for all closed smooth paths C.

ANSWER:

(d) is true since gradient fields are path-independent, by the Fundamental Theorem of Calculus for Line Integrals. This contradicts (c), so that can't be true. As for (a) and (b), the integral of f along any path is the difference between the value of f at the ending point and the value of f at the starting point. Since the difference of two positive numbers could be any number, neither (a) nor (b) is true always.

COMMENT:

Follow-up Question. Under what conditions on f is (b) true? Answer: when f is constant.

ConcepTests and Answers and Comments for Section 18.4 ▬▬▬▬▬▬

1. Figure 18.7 shows a curve C broken into two pieces C_1 and C_2. Which of the following statements is true for any smooth vector field $\vec{F}$? (There might be more than one.)

   (a) $\int_C \vec{F} \cdot d\vec{r} = \int_{C_1} \vec{F} \cdot d\vec{r} + \int_{C_2} \vec{F} \cdot d\vec{r}$

   (b) $\int_C \vec{F} \cdot d\vec{r} = \int_{C_1} \vec{F} \cdot d\vec{r} - \int_{C_2} \vec{F} \cdot d\vec{r}$

   (c) $\int_{C_1} \vec{F} \cdot d\vec{r} = \int_{C_2} \vec{F} \cdot d\vec{r}$.

   (d) $\int_C \vec{F} \cdot d\vec{r} = 0$

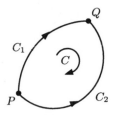

Figure 18.7

   ANSWER:

   Only statement (b) is true for an arbitrary vector field.

   COMMENT:

   Ask the students what conditions on the field are necessary to make statements (c) and (d) true. (E.g., $\vec{F}$ is a gradient field or a path-independent field.)

2. A smooth two dimensional vector field $\vec{F} = F_1 \vec{i} + F_2 \vec{j}$, with $\vec{F} \neq \vec{0}$ satisfies $\dfrac{\partial F_2}{\partial x} = \dfrac{\partial F_1}{\partial y}$ at every point in the plane. Which of the following statements is true? (There might be more than one.)

   (a) $\int_C \vec{F} \cdot d\vec{r} = 0$ for all smooth paths C.

   (b) $\int_{C_1} \vec{F} \cdot d\vec{r} = \int_{C_2} \vec{F} \cdot d\vec{r}$ for any two smooth paths C_1 and C_2 with the same starting and ending points.

   (c) $\vec{F} = \text{grad } f$ for some function f.

   (d) If C_1 is the straight line from -1 to 1 on the y-axis and if C_2 is the right half of the unit circle, traversed counter-clockwise, then $\int_{C_1} \vec{F} \cdot d\vec{r} = \int_{C_2} \vec{F} \cdot d\vec{r}$.

   ANSWER:

   Statements (b), (c), and (d) are true. Green's theorem tells us that $\vec{F}$ is path-independent, and path-independent fields are gradient fields, so statements (b) and (c) are true. Statement (d) is just a particular case of statement (b). Statement (a) would be true of it had the extra condition that C was a closed path, but it is not true for all paths.

   COMMENT:

   Follow-up Question. Would any of the statement be true if the vector field satisfied the partial derivative condition everywhere except at the point $(-1, 0)$? (For example, if it were not defined at that point.) Only statement (d) would be true in this case, since the region enclosed by the two paths does not contain $(-1, 0)$.

3. A vector field $\vec{F}$ in the plane is defined everywhere except at the point $(2, 1)$, and $\dfrac{\partial F_2}{\partial x} = \dfrac{\partial F_1}{\partial y}$ at every point in its domain. For each path C in Figure 18.8, say whether Green's theorem allows you to conclude that $\int_C \vec{F} \cdot d\vec{r} = 0$.

(a)
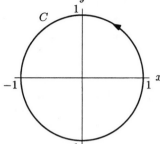

(b)
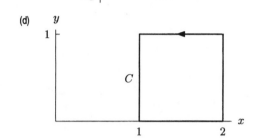

(c)
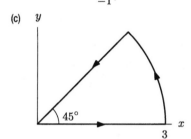

(d)

Figure 18.8

ANSWER:

(a) and (b). For Green's theorem to apply, the region enclosed by the curve cannot contain any holes in the domain of $\vec{F}$. The point $(2, 1)$ is contained in the region enclosed by the path in (c), and is on the path in (d).

COMMENT:

Vary the hypothesis by replacing $(2,1)$ with different points.

Chapter Nineteen

ConcepTests and Answers and Comments for Section 19.1

1. Consider the flux of $\vec{F} = x\vec{i}$ through a disk of radius 1 oriented as in (a)–(d). In which cases is the flux positive?

   (a) In the yz-plane, centered at the origin and oriented in the direction of increasing x.
   (b) In the plane $x = 2$, centered on the x-axis and oriented away from the origin.
   (c) In the plane $y = 2$, centered on the y-axis and oriented away from the origin.
   (d) In the plane $x + y = 2$, centered on the x-axis and oriented away from the origin.

   ANSWER:
   The flux through the disks in parts (b) and (d) are positive. The flux in part (a) is zero, since $\vec{F} = \vec{0}$ on the yz-plane. The flux in part (c) is zero, since $\vec{F}$ is parallel to the disk in this case.

2. Consider the flux of $\vec{F} = y\vec{i}$ through a disk of radius 1 oriented as in (a)–(d). In which cases is the flux positive?

   (a) In the yz-plane, centered at the origin and oriented in the direction of increasing x.
   (b) In the plane $x = 2$, centered on the x-axis and oriented away from the origin.
   (c) In the plane $y = 2$, centered on the y-axis and oriented away from the origin.
   (d) In the plane $x + y = 2$, centered on the x-axis and oriented away from the origin.

   ANSWER:
   None. The flux through the surfaces in parts (a)–(d) is zero. In parts (a), (b), and (d), the flux integral is zero because contributions from positive and negative y-values cancel. The flux in part (c) is zero because the vector field is parallel to the disk.

3. For each oriented surface (I)–(V), pick all the vector fields from (a)–(e) that give a positive flux integral $\int_S \vec{F} \cdot d\vec{A}$.

   (a) $\vec{F} = x\vec{j}$ (b) $\vec{F} = x\vec{k}$ (c) $\vec{F} = -y\vec{k}$
   (d) $\vec{F} = y\vec{j}$ (e) $\vec{F} = -z\vec{i}$ (f) $\vec{F} = (z+x)\vec{i}$

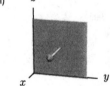

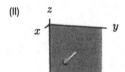

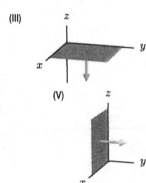

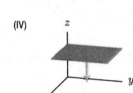

   ANSWER:
   For (I), $\vec{A}$ is in the same direction as $\vec{i}$, the possible answers are (e) and (f). The surface has $x = 0$, $y \geq 0$, $z \geq 0$, so (f) is the only answer.
   For (II), $\vec{A}$ is in the same direction as $\vec{i}$, so the possible answers are (e) and (f). The surface has $x = 0$, $y \geq 0$, $z \leq 0$, so (e) is the only answer
   For (III), $\vec{A}$ is in the same direction as $-\vec{k}$, so possible answers are (b) and (c). The surface has $x \geq 0$, $y \geq 0$, $z = 0$, so (c) is the only answer.
   For (IV), $\vec{A}$ is in the same direction as $-\vec{k}$, so the possible answers are (b) and (c). The surface has $x \geq 0$, $y \geq 0$, $z \geq 0$, so (c) is the only answer.
   For (V), $\vec{A}$ is in the same direction as $\vec{j}$, so the possible answers are (a) and (d). On the surface, $x \geq 0$, $y = 0$, $z \geq 0$, so (a) is the only answer

4. Let $\vec{F} = \vec{r} = x\vec{i} + y\vec{j} + z\vec{k}$. For which of the surfaces S, in (a)–(d) is the flux $\int_S \vec{F} \cdot d\vec{A}$ positive?

   (a) Sphere of radius 1 centered at the origin.
   (b) Unit disk in the xy-plane, oriented upward.
   (c) Unit disk in the plane $x = 2$, oriented toward the origin.
   (d) A cube of side 2, centered at the origin and with sides parallel to the axes.

   ANSWER:

   The flux in parts (a) and (d) are positive. In each case, the surface is oriented away from the origin, and the vector field points outward. The flux in part (b) is zero as the vector field is parallel to the surface. The flux in part (c) is negative as the surface is oriented toward the origin and the vector field points in the direction of increasing x.

   COMMENT:

   As an extension, ask students to compute the flux through these surfaces. Answers are

   (a) 4π (b) 0 (c) -2π (d) 24

5. For each of the surfaces in (a)–(e), pick the vector field $\vec{F}_1$, $\vec{F}_2$, $\vec{F}_3$, $\vec{F}_4$, $\vec{F}_5$, with the largest flux through the surface. The surfaces are all squares of the same size. Note that the orientation is shown.

$$\vec{F}_1 = 2\vec{i} - 3\vec{j} - 4\vec{k}$$
$$\vec{F}_2 = \vec{i} - 2\vec{j} + 7\vec{k}$$
$$\vec{F}_3 = -7\vec{i} + 5\vec{j} + 6\vec{k}$$
$$\vec{F}_4 = -11\vec{i} + 4\vec{j} - 5\vec{k}$$
$$\vec{F}_5 = -5\vec{i} + 3\vec{j} + 5\vec{k}$$

(a)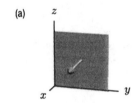

(b)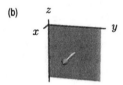

(c)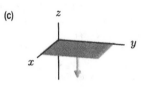

(d)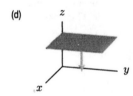

(e)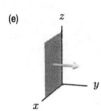

   ANSWER:

   For (a), we want the vector field with the largest $\vec{i}$ component so $\vec{F}_1$.
   For (b), we want the vector field with the largest $\vec{i}$ component so $\vec{F}_1$.
   For (c), we want the vector field with the most negative $\vec{k}$ component, so $\vec{F}_4$.
   For (d), we want the vector field with the most negative $\vec{k}$ component, so $\vec{F}_4$.
   For (e), we want the vector field with the largest $\vec{j}$ component so $\vec{F}_3$.

   COMMENT:

   Assume each surface is a square of side 1. Calculate the largest flux through each surface.

6. Let $\int_S \vec{F}_i \cdot d\vec{A}$ be the flux of the vector field $\vec{F}_i$, for $i = 1$–4, out of the sphere of radius 2 centered at the origin. For the following vector fields, arrange the flux integrals in ascending order.

$$\vec{F}_1 = \frac{\vec{r}}{||\vec{r}||}, \quad \vec{F}_2 = \frac{\vec{r}}{||\vec{r}||^2}, \quad \vec{F}_3 = x\vec{j}, \quad \vec{F}_4 = \vec{r}\,||\vec{r}||.$$

ANSWER:

On the sphere, the vector fields $\vec{F}_1, \vec{F}_2, \vec{F}_4$ point radially outward and are everywhere perpendicular to the surface. Since the radius of the sphere is 2, $||\vec{F}_4|| = 4$, $||\vec{F}_2|| = 1/2$, and $||\vec{F}_1|| = 1$. The vector field $\vec{F}_3$ points inward where $x < 0$ and outward where $x > 0$, and by symmetry, $\int_S \vec{F}_3 \cdot d\vec{A} = 0$. All the other integrals are positive. Thus,

$$\int_S \vec{F}_3 \cdot d\vec{A} < \int_S \vec{F}_2 \cdot d\vec{A} < \int_S \vec{F}_1 \cdot d\vec{A} < \int_S \vec{F}_4 \cdot d\vec{A}.$$

7. Without evaluating them, match the flux integrals (I)–(VI) with the values (a)–(f). The surface B_1 is the closed box, $-1 \leq x \leq 1, -1 \leq y \leq 1, -1 \leq z \leq 1$. The surface B_2 is the closed box, $-2 \leq x \leq 2, -2 \leq y \leq 2, -2 \leq z \leq 2$.

I. $\displaystyle\int_{B_1} \vec{r} \cdot d\vec{A}$ II. $\displaystyle\int_{B_2} \vec{r} \cdot d\vec{A}$ III. $\displaystyle\int_{B_1} \left(\vec{r} - \frac{\vec{r}}{||\vec{r}||^3} \right) \cdot d\vec{A}$

IV. $\displaystyle\int_{B_1} -\frac{\vec{r}}{||\vec{r}||^3} \cdot d\vec{A}$ V. $\displaystyle\int_{B_2} -\vec{r} \cdot d\vec{A}$ VI. $\displaystyle\int_{B_1} -\vec{r} \cdot d\vec{A}$

(a) -24 (b) -12.6 (c) 192 (d) -192 (e) 11.4 (f) 24

ANSWER:

Before matching integrals and answers, we arrange the integrals in ascending order. Since the boxes are closed, they are oriented outward. Thus, the integrals I and II are positive and the integrals IV–VI are negative. Integrals I and II have the same vector field, which increases in magnitude with distance from the origin. Since the surface B_1 in I lies inside the surface B_2 in II, the integral II is larger than I. The vector field $\vec{r} - \vec{r}/||\vec{r}||^3$ in III points outward on the surface of B_1, because $||\vec{r}||$ is greater than or equal to 1 there, so integral III is also positive. Since the vector field in $\vec{r} - \vec{r}/||\vec{r}||^3$ is smaller in magnitude than $\vec{r}$, integral I is larger than III (the surfaces are the same). Thus

$$I = 24\,(f), \quad II = 192\,(c), \quad III = 11.4\,(e).$$

The integrals IV, V, and (VI) are all negative. Since $\vec{r}/||\vec{r}||^3$ is smaller in magnitude than $\vec{r}$, integral IV is smaller in magnitude than VI (the surfaces are the same). Since surface B_1 in VI lies inside B_2 in V, integral VI is smaller in magnitude than integral V. Thus

$$IV = -12.6\,(b), \quad V = -192\,(d), \quad VI = -24\,(a).$$

8. Are the following statements about the flux integral $\int_S \vec{F} \cdot d\vec{A}$ of a vector field $\vec{F}$ through an oriented surface S true or false? If at all points of S the vector field $\vec{F}$ is

(a) Tangent to S, then $\int_S \vec{F} \cdot d\vec{A} = 0$.

(b) Pointing to the same side of S as the orientation normal vector $\vec{n}$, then $\int_S \vec{F} \cdot d\vec{A} > 0$.

(c) Normal to S, then $\int_S \vec{F} \cdot d\vec{A} \neq 0$.

   ANSWER:

   The flux integrals $\int_S \vec{F} \cdot d\vec{A}$ are approximated by Riemann sums

$$\Sigma \vec{F} \cdot \Delta \vec{A} = \Sigma \|\vec{F}\| \|\Delta \vec{A}\| \cos\theta$$

where θ is the angle between the vector field $\vec{F}$ and a surface normal vector $\Delta \vec{A}$ in the direction of the orientation normal vector $\vec{n}$.

(a) True. A surface tangent vector is at right angles to the surface normal vector $\Delta \vec{A}$. In the limit, all dot product terms in the Riemann sums for $\int_S \vec{F} \cdot d\vec{A}$ are zero because the angle θ is a right angle, giving $\cos\theta = 0$. So the flux integral equals zero.

(b) True. A vector at a point of S pointing to the same side of S as the normal vector $\vec{n}$ makes an acute angle with $\vec{n}$. All dot product terms in the Riemann sums for $\int_S \vec{F} \cdot d\vec{A}$ are positive because the angles θ are acute, giving $\cos\theta > 0$. Thus the integral is positive.

(c) False. A counterexample is given by $\vec{F} = x\vec{k}$, and the square surface S in the xy-plane given by $-10 \leq x \leq 10$, $-10 \leq y \leq 10$, oriented upward in the direction of $\vec{k}$. The vector field $\vec{F}$ is everywhere normal to S, but the flux integral $\int_S \vec{F} \cdot d\vec{A}$ is zero because the positive flux through the $x > 0$ part of S cancels by symmetry the negative flux through the $x < 0$ part of S.

ConcepTests and Answers and Comments for Section 19.2

1. The vector field, $\vec{F}$, in Figure 19.1 depends only on z; that is, it is of the form $g(z)\vec{k}$, where g is an increasing function. The integral $\int_S \vec{F} \cdot d\vec{A}$ represents the flux of $\vec{F}$ through this rectangle, S, oriented upward. In each of the following cases, how does the flux change?

   (a) The rectangle is twice as wide in the x-direction, with new corners at the origin, $(2, 0, 0)$, $(2, 1, 3)$, $(0, 1, 3)$.
   (b) The rectangle is moved so that its corners are at $(1, 0, 0)$, $(2, 0, 0)$, $(2, 1, 3)$, $(1, 1, 3)$.
   (c) The orientation is changed to downward.
   (d) The rectangle is tripled in size, so that its new corners are at the origin, $(3, 0, 0)$, $(3, 3, 9)$, $(0, 3, 9)$.

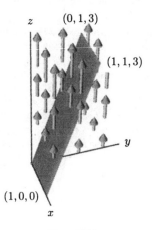

Figure 19.1

ANSWER:

   (a) The surface can be considered as made up of two rectangles, the original one and another "next door" along the x-axis. Because the vector field is independent of x, the flux through both rectangles are the same. Thus, the flux has doubled.
   (b) Because the vector field does not depend on x, the flux is unchanged.
   (c) The sign is reversed, but the magnitude of the flux remains the same.
   (d) Tripling each side of the rectangle multiplies its area by 9. However, the surface now extends further up the z-axis, where the vector field is not given. If the vector field is larger further up the z-direction (as suggested by the diagram), then the flux has multiplied by a factor of more than 9.

2. To calculate the flux through each of the following surfaces, what coordinate system, rectangular, cylindrical, or spherical, would be a good choice?

   (a) $x^2 + y^2 = 9$
   (b) $x - y + z = 5$
   (c) $9 - x^2 - y^2 - z^2 = 0$
   (d) $y^2 + z^2 = 5$

   ANSWER:

   (a) Cylindrical, because the surface is a cylinder.
   (b) Rectangular, because the surface is a plane which can be written as the graph of a function of x and y.
   (c) Spherical, because the surface is a sphere.
   (d) Cylindrical, because the surface is a cylinder. Since the cylinder is centered on the x-axis, we will have to use x in place of z.

3. To calculate a flux integral through the surfaces (I)–(V), which area element (a)–(e) would be a good choice? (There may be more than one solution.) If you choose (f), give the function f you would use.

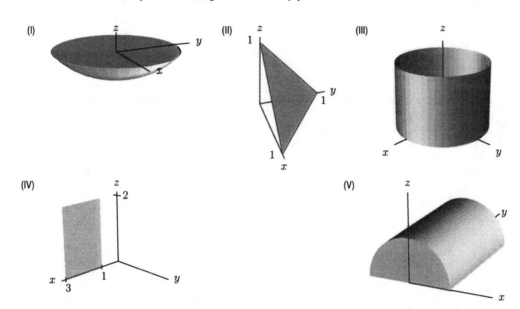

   (a) $d\vec{A} = \vec{k}\, dx\, dy$
   (b) $d\vec{A} = \vec{j}\, dx\, dz$
   (c) $d\vec{A} = \vec{i}\, dy\, dz$
   (d) $d\vec{A} = (\sin\phi\cos\theta\vec{i} + \sin\phi\sin\theta\vec{j} + \cos\phi\vec{k})R^2\sin\phi\, d\phi\, d\theta$
   (e) $d\vec{A} = (\cos\theta\vec{i} + \sin\theta\vec{j})R\, dz\, d\theta$
   (f) $d\vec{A} = (-f_x\vec{i} + f_y\vec{j} + \vec{k})\, dx\, dy$

   ANSWER:

   (I) For part of a sphere, use spherical coordinates, (d), or rectangular, (f) with $f(x,y) = -\sqrt{r^2 - x^2 - y^2}$.
   (II) Use rectangular coordinates, (f), with $f(x,y) = c + mx + ny$.
   (III) Use cylindrical coordinates, (e).
   (IV) Since the surface is in the xz-plane, use (b).
   (V) For the half cylinder, we could use cylindrical coordinates (centered on the y-axis rather than the z-axis), or rectangular, (f) with $z = \sqrt{r^2 - x^2}$.

ConcepTests and Answers and Comments for Section 19.3 ──────────

1. An oriented surface S is parameterized by $\vec{r} = \vec{r}(s,t)$. The flux of a vector field $\vec{F}$ through S is given by

$$Q = \int_R \vec{F}(\vec{r}(s,t)) \cdot \left(\frac{\partial \vec{r}}{\partial s} \times \frac{\partial \vec{r}}{\partial t}\right) \, ds \, dt.$$

In each of the following cases, how is the value of the new integral related to Q?

(a) $\displaystyle\int_R \vec{F}(\vec{r}(s,t)) \cdot \left(\frac{\partial \vec{r}}{\partial t} \times \frac{\partial \vec{r}}{\partial s}\right) \, ds \, dt$

(b) $\displaystyle\int_R 2\vec{F}(\vec{r}(s,t)) \cdot \left(2\frac{\partial \vec{r}}{\partial s}\right) \times \left(2\frac{\partial \vec{r}}{\partial t}\right) \, ds \, dt$

(c) $\displaystyle\int_R \left(\frac{\partial \vec{r}}{\partial s} \times \frac{\partial \vec{r}}{\partial t}\right) \cdot \vec{F}(\vec{r}(s,t)) \, ds \, dt$

(d) $\displaystyle\int_R \|\vec{F}(\vec{r}(s,t))\| \left\|\frac{\partial \vec{r}}{\partial s}\right\| \left\|\frac{\partial \vec{r}}{\partial t}\right\| \, ds \, dt$

ANSWER:

(a) Since the order of the factors in the cross product has been reversed, the orientation of the surface has been reversed. Thus, the integral is $-Q$.

(b) Since each factor in the integral has been multiplied by 2, the integral is $8Q$.

(c) The order of the vectors in a dot product does not affect the answer, so this integral is also Q.

(d) Since a dot product of two vectors is no more than the product of the lengths of the vectors, and the length of a cross product is also less than the product of the lengths of the two vectors, this integral is certainly positive, and at least as large as $|Q|$.

2. Let S be the hemisphere $x^2 + y^2 + z^2 = 1$ with $x \leq 0$, oriented away from the origin. Which of the following integrals represents the flux of $\vec{F}(x,y,z)$ through S?

(a) $\displaystyle\int_R \vec{F}(x,y,z(x,y)) \cdot \frac{\partial \vec{r}}{\partial x} \times \frac{\partial \vec{r}}{\partial y} \, dx \, dy$

(b) $\displaystyle\int_R \vec{F}(x,y,z(x,y)) \cdot \frac{\partial \vec{r}}{\partial y} \times \frac{\partial \vec{r}}{\partial x} \, dy \, dx$

(c) $\displaystyle\int_R \vec{F}(x,y(x,z),z) \cdot \frac{\partial \vec{r}}{\partial x} \times \frac{\partial \vec{r}}{\partial z} \, dx \, dz$

(d) $\displaystyle\int_R \vec{F}(x,y(x,z),z) \cdot \frac{\partial \vec{r}}{\partial z} \times \frac{\partial \vec{r}}{\partial x} \, dz \, dx$

(e) $\displaystyle\int_R \vec{F}(x(y,z),y,z) \cdot \frac{\partial \vec{r}}{\partial y} \times \frac{\partial \vec{r}}{\partial z} \, dy \, dz$

(f) $\displaystyle\int_R \vec{F}(x(y,z),y,z) \cdot \frac{\partial \vec{r}}{\partial z} \times \frac{\partial \vec{r}}{\partial y} \, dz \, dy$

ANSWER:

Since the half sphere is $x = -\sqrt{1 - y^2 - z^2}$, we parameterize in the form $x = x(y,z)$. Thus, the answer is either (e) or (f). We have $\partial \vec{r}/\partial y = (y/\sqrt{1 - y^2 - z^2})\vec{i} + \vec{j}$ and $\partial \vec{r}/\partial z = (z/\sqrt{1 - y^2 - z^2})\vec{i} + \vec{k}$, and $\partial \vec{r}/\partial y \times \partial \vec{r}/\partial z = \vec{i} - (y/\sqrt{1 - y^2 - z^2})\vec{j} - (z/\sqrt{1 - y^2 - z^2})\vec{k} = -\vec{r}/\sqrt{1 - y^2 - z^2}$. Our surface is oriented away from the origin, so we want $(\partial \vec{r}/\partial z) \times (\partial \vec{r}/\partial y) = \vec{r}/\sqrt{1 - y^2 - z^2}$, so the answer is (f).

Chapter Twenty

ConcepTests and Answers and Comments for Section 20.1

1. If $\vec{F}(x, y, z)$ is a vector field and $f(x, y, z)$ is a scalar function, which of the following are **not** defined?

   (a) grad $\vec{F}$ (b) grad f (c) div $\vec{F} + f$
   (d) $\vec{F} + $ grad f (e) div $\vec{F} + $ grad f

   ANSWER:

   (a) Not defined
   (b) Defined
   (c) Defined
   (d) Defined
   (e) Not defined; can't add vector and scalar

2. Which of the following quantities are vectors?

   (a) div$(\vec{F}(x, y, z))$ (b) grad $f(x, y, z) \cdot \vec{u}$ (c) div grad$(f(x, y, z))$
   (d) $(\text{div } \vec{F}(x, y, z))\vec{F}$ (e) $f(x, y, z)$ grad $f(x, y, z)$

   ANSWER:

   Only (d) and (e) are vectors.

3. For $\vec{r} = x\vec{i} + y\vec{j} + z\vec{k}$, an arbitrary function $f(x, y, z)$ and an arbitrary vector field, $\vec{F}(x, y, z)$, which of the following is a vector field, and which is a constant vector field?

   (a) grad f (b) $(\text{div } \vec{F})\vec{i}$ (c) $(\text{div } \vec{r})\vec{i}$ (d) $(\text{div } \vec{i})\vec{F}$ (e) grad(div $\vec{F}$)

   ANSWER:

   (a), (b), and (e) all depend on the point (x, y, z), so they are vector fields. Since div $\vec{r} = 3$ and div $\vec{i} = 0$, the vectors in (c) and (d) are constant vector fields.

4. Are the following statements true or false? If all the flow lines of a vector field $\vec{F}$

   (a) Are parallel straight lines, then div $\vec{F} = 0$.
   (b) Radiate outward along straight lines from the origin, then div $\vec{F} > 0$.

   ANSWER:

   (a) False. A counterexample is given by $\vec{F} = x\vec{i}$. The flow lines of $\vec{F}$ are all straight lines parallel to the vector $\vec{i}$, but div $\vec{F} = 1 \neq 0$. Imagine cars driving on a straight highway where each car is always going slower than the car just ahead of it. The separation between the cars is increasing, so the cars are diverging.
   (b) False. A counterexample is given by $\vec{F} = (x^2 + y^2 + z^2)^{-2}(x\vec{i} + y\vec{j} + z\vec{k})$ in which the speed of the flow rapidly decreases as it moves away from the origin. For this vector field div $\vec{F} = -(x^2 + y^2 + z^2)^{-2} < 0$.

   This problem illustrates the fact that divergence can be caused by expansion and compression in the direction of the flow of a vector field, not only by convergence and divergence of the flow lines.

ConcepTests and Answers and Comments for Section 20.2

1. Let $\vec{F} = (5x + 7y)\vec{i} + (7y + 9z)\vec{j} + (9z + 11x)\vec{k}$, and let Q_i be the flux of $\vec{F}$ through the surfaces S_i for $i = 1\text{–}4$. Arrange Q_i in ascending order, where

   (a) S_1 is the sphere of radius 2 centered at the origin
   (b) S_2 is the cube of side 2 centered at the origin and with sides parallel to the axes
   (c) S_3 is the sphere of radius 1 centered at the origin
   (d) S_4 is a pyramid with all four corners lying on S_3

   ANSWER:

   Since div $\vec{F} = 5 + 7 + 9 = 21$, by the Divergence Theorem, if W_i is the region inside S_i, we have

   $$Q_i = \int_{W_i} 21 \, dV = 21 \cdot \text{Vol } W_i.$$

   Thus, we arrange Q_i by the volume of W_i, so

   $$Q_4 < Q_3 < Q_2 < Q_1.$$

2. Let S_i be the flux vector field $\vec{F}_i$ out of the unit sphere centered at the origin, for $i = 1$–4. Arrange S_i in ascending order, where

$$\vec{F}_1 = (e^z + x^3)\vec{i} + e^x\vec{j} + y^3\vec{k}$$
$$\vec{F}_2 = (z^3 + \cos y)\vec{i} - y^3\vec{j} + x^3 y^3\vec{k}$$
$$\vec{F}_3 = z^2\vec{i} - (x^2 + z^2)\vec{j} + (z^3 + zy^2)\vec{k}$$
$$\vec{F}_4 = (x^4 - y^4)\vec{i} + (z^4 - 2x^3 y)\vec{j} + (y^4 - 2x^3 z)\vec{k}$$

ANSWER:

We have

$$\text{div } \vec{F}_1 = 3x^2$$
$$\text{div } \vec{F}_2 = -3y^2$$
$$\text{div } \vec{F}_3 = 3z^2 + y^2$$
$$\text{div } \vec{F}_4 = 4x^3 - 2x^3 - 2x^3 = 0.$$

By the Divergence Theorem, if W is the region inside the sphere

$$S_1 = \int_W 3x^2\,dV, \quad S_2 = -\int_W 3y^2\,dV, \quad S_3 = \int_W (3z^2 + y^2)\,dV, \quad S_4 = \int_W 0\,dV = 0.$$

Since the region of integration, W, is the same in each integral, we have

$$S_2 < S_4 = 0 < S_1 < S_3.$$

3. For surface S_i, with $i = 1$–4, the flux of $\vec{F} = \vec{r}/||\vec{r}||^3$ out of S_i is Q_i. Which of the Q_i are equal to one another? What are their values?

$$S_1 : \text{Sphere } (x - 3)^2 + (y - 3)^2 + z^2 = 1$$
$$S_2 : \text{Sphere } x^2 + (y - 1)^2 + z^2 = 3$$
$$S_3 : \text{Box } x = \pm 2, y = \pm 1, z = \pm 2$$
$$S_4 : \text{Box } x = 1, x = 3, y = \pm 1, z = \pm 1$$
$$S_5 : \text{Sphere of radius 2 centered at the origin}$$
$$S_6 : \text{Ellipsoid } x^2/4 + y^2 + 3z^2 = 1.$$

ANSWER:

We know that $\text{div } \vec{F} = 0$ except at the origin. If the surface encloses the origin, the flux is 4π, otherwise, the flux is 0. So

$$Q_1 = Q_4 = 0$$
$$Q_2 = Q_3 = Q_5 = Q_6 = 4\pi.$$

4. Are the following statements true or false? The vector field $\vec{F}$ is defined everywhere in a region W bounded by a surface S.

   (a) If div $\vec{F} > 0$ at all points of W, then the vector field $\vec{F}$ points outward at all points of S.
   (b) If div $\vec{F} > 0$ at all points of W, then the vector field $\vec{F}$ points outward at some points of S.
   (c) If $\int_S \vec{F} \cdot d\vec{A} > 0$, then div $\vec{F} > 0$ at some points of W.

   ANSWER:

   (a) False. A counterexample is given by $\vec{F} = x\vec{i}$, with W the solid cube $1 \le x, y, z \le 2$, and S the boundary surface of W. We have div $\vec{F} = 1 > 0$. But $\vec{F}$ points inward on the square face of S where $x = 1$, outward on the face where $x = 2$, and tangent to the other four faces.
   (b) True. By the Divergence Theorem, $\int_W \text{div} \, \vec{F} \, dV = \int_S \vec{F} \cdot d\vec{A}$. Since div $\vec{F} > 0$ at all points of W, we have $\int_W \text{div} \, \vec{F} \, dV > 0$. This means that $\int_S \vec{F} \cdot d\vec{A} > 0$, which is impossible if $\vec{F}$ is tangent to S or points inward at every point of S. Thus $\vec{F}$ points outward at some point of S.
   (c) True. By the Divergence Theorem, $\int_W \text{div} \, \vec{F} \, dV = \int_S \vec{F} \cdot d\vec{A} > 0$. The integral $\int_W \text{div} \, \vec{F} \, dV$ can not be positive if div $\vec{F}$ is zero or negative at all points of W. Thus div $\vec{F}$ is positive at some point of W.

ConcepTests and Answers and Comments for Section 20.3

1. Of the following vector fields, which ones have a curl which is parallel to one of the axes? Which axis?

   (a) $y\vec{i} - x\vec{j} + z\vec{k}$ (b) $y\vec{i} + z\vec{j} + x\vec{k}$ (c) $-z\vec{i} + y\vec{j} + x\vec{k}$
   (d) $x\vec{i} + z\vec{j} - y\vec{k}$ (e) $z\vec{i} + x\vec{j} + y\vec{k}$

   ANSWER:

   (a)
   $$\text{curl}(y\vec{i} - x\vec{j} + z\vec{k}) = \begin{vmatrix} \vec{i} & \vec{j} & \vec{k} \\ \frac{\partial}{\partial x} & \frac{\partial}{\partial y} & \frac{\partial}{\partial z} \\ y & -x & z \end{vmatrix} = -2\vec{k}$$

   (b)
   $$\text{curl}(y\vec{i} + z\vec{j} + x\vec{k}) = \begin{vmatrix} \vec{i} & \vec{j} & \vec{k} \\ \frac{\partial}{\partial x} & \frac{\partial}{\partial y} & \frac{\partial}{\partial z} \\ y & z & x \end{vmatrix} = -\vec{i} - \vec{j} - \vec{k}$$

   (c)
   $$\text{curl}(-z\vec{i} + y\vec{j} + x\vec{k}) = \begin{vmatrix} \vec{i} & \vec{j} & \vec{k} \\ \frac{\partial}{\partial x} & \frac{\partial}{\partial y} & \frac{\partial}{\partial z} \\ -z & y & x \end{vmatrix} = -2\vec{j}$$

   (d)
   $$\text{curl}(x\vec{i} + z\vec{j} - y\vec{k}) = \begin{vmatrix} \vec{i} & \vec{j} & \vec{k} \\ \frac{\partial}{\partial x} & \frac{\partial}{\partial y} & \frac{\partial}{\partial z} \\ x & z & -y \end{vmatrix} = -2\vec{i}$$

   (e)
   $$\text{curl}(z\vec{i} + x\vec{j} + y\vec{k}) = \begin{vmatrix} \vec{i} & \vec{j} & \vec{k} \\ \frac{\partial}{\partial x} & \frac{\partial}{\partial y} & \frac{\partial}{\partial z} \\ z & x & y \end{vmatrix} = \vec{i} + \vec{j} + \vec{k}$$

   So (a), (c), (d) are parallel to the z-, y-, and x-axes, respectively.

2. Let $\vec{F}(x, y, z)$ be a vector field and let $f(x, y, z)$ be a scalar function, both defined for all (x, y, z). If $\vec{r} = x\vec{i} + y\vec{j} + z\vec{k}$, which of the following are not defined?

(a) curl f (b) grad $\vec{F}$ (c) curl $\vec{F}$ + grad f

(d) curl($\vec{r} \times$ grad f)   (e) $f +$ div $\vec{F}$

ANSWER:

(a) Not defined; we find the curl of a vector field.
(b) Not defined; we find the gradient of a scalar function.
(c) Defined; this is the sum of two vector fields.
(d) Defined; this is the curl of the vector field $\vec{r} \times$ grad f.
(e) Defined; this is the sum of two scalar functions.

3. Are the following statements true or false? If all the flow lines of a vector field $\vec{F}$

(a) Are straight lines, then curl $\vec{F} = \vec{0}$.
(b) Lie in planes parallel to the xy-plane, then curl $\vec{F}$ is a multiple of $\vec{k}$ at every point.
(c) Go around the z axis in circles centered on the z-axis and parallel to the xy-plane, then curl $\vec{F}$ is a multiple of $\vec{k}$ at every point.

ANSWER:

(a) False. A counterexample is given by $\vec{F} = y\vec{i}$. The flow lines of $\vec{F}$ are all straight lines parallel to the vector $\vec{i}$, but curl $\vec{F} = -\vec{k} \neq \vec{0}$. Imagine a straight river where water moves downstream faster near the left bank than near the right bank. The difference in water speed causes an object in the river to spin to the right as it floats straight downstream. This spinning is an indication of curl, even though the flow is straight.
(b) False. A counterexample is given by $\vec{F} = z\vec{i}$. The flow lines of $\vec{F}$ are all straight lines parallel to the vector $\vec{i}$ and hence all parallel to the xy-plane, but curl $\vec{F} = \vec{j}$.
(c) False. A counterexample is given by $\vec{F} = z(-y\vec{i} + x\vec{j})$. The flow of the vector field $\vec{F}$ is parallel to the xy-plane because $\vec{F}$ has no $\vec{k}$ component. In the plane $z = B$, flows are given by the parametrized circles $x = A\cos Bt, y = A\sin Bt, z = B$ centered on the z-axis. But curl $\vec{F} = -x\vec{i} - y\vec{j} + 2z\vec{k}$ is not parallel to $\vec{k}$ except on the z-axis.

This problem illustrates the fact that curl can be caused by shear in the flow of a vector field, not only by curving of the flow.

ConcepTests and Answers and Comments for Section 20.4

1. The circle C has radius 3 and lies in a plane through the origin. Let $\vec{F} = (2z + 3y)\vec{i} + (x - z)\vec{j} + (6y - 7x)\vec{k}$. What is the equation of the plane and what is the orientation of the circle that make the circulation, $\int_C \vec{F} \cdot d\vec{r}$, a maximum? [Note: You should specify the orientation of the circle by saying that it is clockwise or counterclockwise when viewed from the positive or negative x- or y- or z-axis.]

ANSWER:
Since

$$\text{curl } \vec{F} = \begin{vmatrix} \vec{i} & \vec{j} & \vec{k} \\ \frac{\partial}{\partial x} & \frac{\partial}{\partial y} & \frac{\partial}{\partial z} \\ 2z + 3y & x - z & 6y - 7x \end{vmatrix} = (6 - (-1))\vec{i} - (-7 - 2)\vec{j} + (1 - 3)\vec{k} = 7\vec{i} + 9\vec{j} - 2\vec{k},$$

by Stokes' Theorem, if S is the flat interior of the circle, with area vector $\vec{A}$,

$$\int_C \vec{F} \cdot d\vec{r} = \int_S \text{curl } \vec{F} \cdot d\vec{A} = \int_S (7\vec{i} + 9\vec{j} - 2\vec{k}) \cdot d\vec{A} = (7\vec{i} + 9\vec{j} - 2\vec{k}) \cdot \vec{A}.$$

This is a maximum if $\vec{A}$ is parallel to curl $\vec{F} = 7\vec{i} + 9\vec{j} - 2\vec{k}$, so curl $\vec{F}$ is the normal to the plane of the circle. Since the plane containing the circle is through the origin, its equation is

$$7x + 9y - 2z = 0.$$

To give the orientation of S shown by curl $\vec{F}$, notice that curl $\vec{F}$ points downward (because the z-component is negative). Therefore, the circle must be oriented counterclockwise when seen from the negative z-axis, or clockwise from the positive z-axis.

2. Without evaluating them, decide which of the following integrals have the same value.

    (a) $\int_S \operatorname{curl}(y^3\vec{i} + z^3\vec{j} - x^3\vec{k}) \cdot d\vec{A}$ where S is the disk $x^2 + y^2 \leq 2, z = 3$, oriented upward.

    (b) $\int_S \operatorname{curl}(y^3\vec{i} + z^3\vec{j} - x^3\vec{k}) \cdot d\vec{A}$ where S is the top half of the sphere $x^2 + y^2 + (z - 3)^2 = 2$, oriented upward.

    (c) $\int_C (y^3\vec{i} + z^3\vec{j} - x^3\vec{k}) \cdot d\vec{r}$ where C is the circle $x^2 + y^2 = 2, z = 3$, oriented counterclockwise when viewed from above.

    (d) $\int_S \operatorname{curl}(y^3\vec{i} + z^3\vec{j} - x^3\vec{k}) \cdot d\vec{A}$ where S is the disk $y^2 + z^2 \leq 2, x = 3$, oriented away from the origin.

    ANSWER:

    All four integrals are equal. Integrals (a) and (b) are both equal to integral (c) by Stokes' Theorem. We have

    $$\operatorname{curl}(y^3\vec{i} + z^3\vec{j} - x^3\vec{k}) = \begin{vmatrix} \vec{i} & \vec{j} & \vec{k} \\ \frac{\partial}{\partial x} & \frac{\partial}{\partial y} & \frac{\partial}{\partial z} \\ y^3 & z^3 & -x^3 \end{vmatrix} = -3z^2\vec{i} + 3x^2\vec{j} - 3y^2\vec{k}.$$

    In part (a), $d\vec{A} = \vec{k}\, dx\, dy$, so Stokes' Theorem shows that the integrals in (a), (b), (c) are all given by

    $$\text{Integral (a)} = \int_{x^2+y^2\leq 2, z=3} (-3z^2\vec{i} + 3x^2\vec{j} - 3y^2\vec{k}) \cdot \vec{k}\, dx\, dy = \int_{x^2+y^2\leq 2} -3y^2 dx\, dy.$$

    Applying Stokes' Theorem to the integral in (d), where $d\vec{A} = \vec{i}\, dy\, dz$, we have

    $$\text{Integral (d)} = \int_{y^2+z^2\leq 2, x=3} (-3z^2\vec{i} + 3x^2\vec{j} - 3y^2\vec{k}) \cdot \vec{i}\, dy\, dz = \int_{y^2+z^2\leq 2} -3z^2 dy\, dz.$$

    By substituting y for z and x for y in the integral for (d), we see that the integrals for (a) and (d) are equal. Thus, all four integrals have the same value.

    COMMENT:

    Is the common value of the integral positive or negative? (Answer: negative.) Find the value of the integral. (Requires the integral table or numerical methods.)

3. Figure 20.1 shows the vector field $\operatorname{curl}\vec{F}$ at points of a disk; no formula for the vector field $\vec{F}$ is given. The oriented curve C is a circle, perpendicular to $\operatorname{curl}\vec{F}$. The line integral $\int_C \vec{F} \cdot d\vec{r}$ is

    (a) Positive
    (b) Negative
    (c) Zero
    (d) Can't tell without further information.

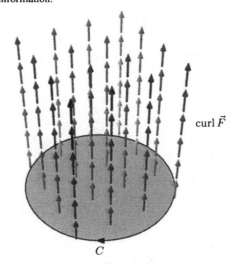

curl $\vec{F}$

C

Figure 20.1

    ANSWER:

    (b) By Stokes' Theorem $\int_C \vec{F} \cdot d\vec{r} = \int_S \operatorname{curl}\vec{F} \cdot d\vec{A}$, where S is the circular disk bounded by C and the surface S is oriented downward, with orientation normal vector $\vec{n}$. Since $\operatorname{curl}\vec{F}$ points to the opposite side of S from $\vec{n}$ at every point of S, the flux integral of $\operatorname{curl}\vec{F}$ through S is negative. Thus, the line integral of $\vec{F}$ over C is also negative.

4. Figure 20.2 shows the vector field $\vec{F}$ at points of a disk. The surface S is oriented upward and perpendicular to $\vec{F}$ at every point. The flux of curl $\vec{F}$ through the surface, $\int_S \text{curl}\,\vec{F} \cdot d\vec{A}$ is

(a) Positive
(b) Negative
(c) Zero
(d) Can't tell without further information

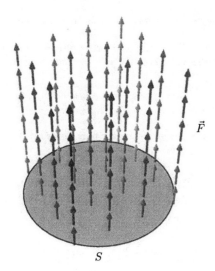

$\vec{F}$

S

Figure 20.2

ANSWER:
(c) By Stokes' theorem $\int_S \text{curl}\,\vec{F} \cdot d\vec{A} = \int_C \vec{F} \cdot d\vec{r}$ where C is the oriented path along the boundary (circumference) of the disk S. Since $\vec{F}$ is perpendicular to the direction of C at every point of C, the line integral of $\vec{F}$ over C is zero. Thus, the flux integral of curl $\vec{F}$ through S is also zero.

ConcepTests and Answers and Comments for Section 20.5

1. Let $f(x, y, z)$ be a scalar function with continuous second partial derivatives. Let $\vec{F}(x, y, z)$ be a vector field with continuous second partial derivatives. Which of the following quantities are identically zero?

(a) curl grad f (b) $\vec{F} \times \text{curl}\,\vec{F}$ (c) grad div $\vec{F}$ (d) div curl $\vec{F}$ (e) div grad f

ANSWER:

(a) is zero, since, for example, $\dfrac{\partial}{\partial z}\left(\dfrac{\partial f}{\partial y}\right) - \dfrac{\partial}{\partial y}\left(\dfrac{\partial f}{\partial z}\right) = 0$.

(b) is not zero. Let $\vec{F} = x\vec{j} + x\vec{k}$, then

$$\text{curl}\,\vec{F} = \begin{vmatrix} \vec{i} & \vec{j} & \vec{k} \\ \frac{\partial}{\partial x} & \frac{\partial}{\partial y} & \frac{\partial}{\partial z} \\ 0 & x & x \end{vmatrix} = -\vec{j} + \vec{k},$$

so $\vec{F} \times \text{curl}\,\vec{F} \neq \vec{0}$.

(c) has components like $\dfrac{\partial}{\partial x}\left(\dfrac{\partial F_1}{\partial x} + \dfrac{\partial F_2}{\partial y} + \dfrac{\partial F_3}{\partial z}\right)$, not zero.

(d) is zero, since $\dfrac{\partial}{\partial x}\left(\dfrac{\partial F_3}{\partial y} - \dfrac{\partial F_2}{\partial z}\right) + \dfrac{\partial}{\partial y}\left(\dfrac{\partial F_1}{\partial z} - \dfrac{\partial F_3}{\partial x}\right) + \dfrac{\partial}{\partial z}\left(\dfrac{\partial F_2}{\partial x} - \dfrac{\partial F_1}{\partial y}\right) = 0$.

(e) is $\dfrac{\partial^2 f}{\partial x^2} + \dfrac{\partial^2 f}{\partial y^2} + \dfrac{\partial^2 f}{\partial z^2}$, not zero.

Problems 2–3 concern the following integrals:

$I_1 = \int_S \text{curl}(y^3\vec{i} + z^3\vec{j} + x^3\vec{k}) \cdot d\vec{A}$, where S is the sphere $x^2 + y^2 + (z-4)^2 = 9$.

$I_2 = \int_C \text{grad}(y^3 + z^3 + x^3) \cdot d\vec{r}$, where C is the circle $x^2 + (y-4)^2 = 9, z = 5$, oriented counterclockwise when viewed from above.

$I_3 = \int_W \text{div}(\text{grad}(x^2 + y^2 + z^2)) \, dV$, where W is the interior of the sphere $x^2 + y^2 + (z-4)^2 \leq 9$.

$I_4 = \int_C \text{grad}\,\text{div}(x^3\vec{i} + y^3\vec{j} + z^3\vec{k}) \cdot d\vec{r}$, where C is the line from $(0,0,0)$ to $(1,2,3)$.

2. Without calculating any of them, decide which of the integrals I_1, I_2, I_3, I_4 are 0.

    ANSWER:

    Since the integral of a curl vector field over a closed surface gives 0, we know $I_1 = 0$. Since the integral of a gradient vector field around a closed curve gives 0, we know $I_2 = 0$. The other integrals are not zero

3. Calculate the integrals I_1, I_2, I_3, I_4.

    ANSWER:

    $I_1 = 0$, since the surface is closed.

    $I_2 = 0$, since the curve is closed.

    Since $\text{grad}(x^2 + y^2 + z^2) = 2x\vec{i} + 2y\vec{j} + 2z\vec{k}$, we have $\text{div}\,\text{grad}(x^2 + y^2 + z^3) = 2 + 2 + 2 = 6$, so

    $$I_3 = \int_W 6 \, dV = 6 \cdot \text{Volume of } V = 6 \cdot \frac{4}{3}\pi 3^3 = 216\pi.$$

Since $\text{div}(x^3\vec{i} + y^3\vec{j} + z^3\vec{k}) = 3x^2 + 3y^2 + 3z^2$, by the Fundamental Theorem of Line Integrals, we know that I_4 is the difference in the values of the potential function, $3x^2 + 3y^2 + 3z^2$, at the end points:

$$I_4 = \int_C \text{grad}(3x^2 + 3y^2 + 3z^2) \cdot d\vec{r} = 3x^2 + 3y^2 + 3z^2 \Big|_{(0,0,0)}^{(1,2,3)} = 42.$$